United States
Environmental Protection
Agency

# Literature Review of Contaminants in Livestock and Poultry Manure and Implications for Water Quality

July 2013

Office of Water (4304T)
EPA 820-R-13-002
July 2013

# Acknowledgements and Disclaimer

This document is designed to provide technical background information for the USEPA's Office of Water research efforts. This report makes no policy or regulatory recommendations; it does identify information gaps that may help define research needs for USEPA and its federal, state, and local partners to better understand these issues. The Lead USEPA Scientist is Octavia Conerly and Co-Lead Lesley Vazquez Coriano, Health and Ecological Criteria Division, Office of Science and Technology, Office of Water. This document was prepared under USEPA contract No. GS-10F-0105J, Task Order 1107 with The Cadmus Group, Inc. This report received technical expert reviews from many scientists within USEPA and from the U.S. Department of Agriculture, Agricultural Research Service.

This document is not a regulation or guidance. Mention of trade names or commercial products does not constitute an endorsement or recommendation for use.

This page intentionally left blank.

# Acronyms and Abbreviations

| | |
|---|---|
| AFO | Animal Feeding Operation |
| ARS | Agricultural Research Service |
| AU | Animal Unit |
| AWWA | American Water Works Association |
| BMP | Best Management Practice |
| BOD | Biochemical Oxygen Demand |
| BVDV | Bovine Viral Diarrhea Virus |
| CAFO | Concentrated Animal Feeding Operation |
| CDC | Centers for Disease Control |
| CENR | Committee on Environment and Natural Resources |
| CFR | Code of Federal Regulations |
| CIDR | Controlled Internal Drug Release |
| DNA | Deoxyribonucleic Acid |
| ECOSAR | Ecological Structure Activity Relationships |
| EHEC | Escherichia coli O157:H7 |
| EQIP | Environmental Quality Incentives Program |
| ERS | Economics Research Service |
| GAC | Granular Activated Carbon |
| HAB | Harmful Algal Bloom |
| HEV | Hepatitis E Virus |
| HUS | Hemolytic-Uremic Syndrome |
| MCL | Maximum Contaminant Level |
| NAHMS | National Animal Health Monitoring System |
| NARMS | National Antimicrobial Resistance Monitoring System |
| NAS | National Academy of Sciences |
| NITG | Nutrient Innovations Task Group |
| NPDES | National Pollutant Discharge Elimination System |
| NPS | Nonpoint Source |
| NRC | National Research Council |
| NRCS | Natural Resources Conservation Service |
| NRDC | Natural Resources Defense Council |
| NYSDEC | New York State Department of Environmental Conservation |
| ODTS | Organic Dust Toxic Syndrome |

| | |
|---|---|
| OST | Office of Science and Technology |
| PAC | Powdered Activated Carbon |
| PCIFAP | Pew Commission on Industrial Farm Animal Production |
| PCR | Polymerase Chain Reaction |
| PHAC | Public Health Agency of Canada |
| RNA | Ribonucleic Acid |
| rBGH | Recombinant Bovine Growth Hormone |
| RO | Reverse Osmosis |
| SPARROW | SPAtially Referenced Regressions On Watershed attributes |
| TMDL | Total Maximum Daily Load |
| USDA | United States Department of Agriculture |
| USEPA | United States Environmental Protection Agency |
| USFDA | United States Food and Drug Administration |
| USGAO | United States Government Accountability Office |
| USGS | United States Geological Survey |
| WBDO | Waterborne Disease Outbreak |
| WBDOSS | Waterborne Disease and Outbreak Surveillance System |
| WHO | World Health Organization |
| WRI | World Resources Institute |

# Executive Summary

This *Literature Review of Contaminants in Livestock and Poultry Manure and Implications for Water Quality* was prepared by the United States Environmental Protection Agency (USEPA) as part of ongoing efforts to better understand the environmental occurrence and potential effects related to contaminants of emerging concern. Past reviews of animal manure have focused primarily on nutrient issues. This report focuses on summarizing technical information on other components, particularly pathogens and contaminants of emerging concern such as antimicrobials and hormones that may affect water quality. The report makes no policy or regulatory recommendations; it does identify information gaps that may help define research needs for USEPA and its federal, state and local partners to better understand these issues.

Over the past 60 years in the United States (U.S.), farm operations have become fewer in number but larger in size. This has been particularly true in livestock and poultry production. Since the 1950s, the production of livestock and poultry in the U.S. has more than doubled; however, the number of operations has decreased by 80%. Food animal production has shifted to more concentrated facilities with animals often raised in confinement. Production has also become more regionally concentrated. This has been done, in part, to meet the demands for meat and animal products from a growing human population in the U.S. and abroad.

The U.S. Department of Agriculture's (USDA) 2007 Census of Agriculture data are used to estimate beef and dairy cattle, swine, and poultry production. Using standard USDA methods, an estimated 2.2 billion head of livestock and poultry generated approximately 1.1 billion tons of manure in 2007. Manure can be a valuable resource as a natural fertilizer. However, if not managed properly, manure can degrade environmental quality, particularly surface water and ground water resources. The increasing concentration of animal production can lead to concentrations of manure that exceed the beneficial needs of the farmland where it was produced. A 2001 report from the USDA's Economic Research Service found that 60%-70% of the manure nitrogen and phosphorus may not be able to be assimilated by the farmland on which it was generated. As an example of the increasing concentration of production, from 1997 to 2007, the number of swine produced in the US increased by 45%, but the number of swine farms decreased by 30%; over 40% of all swine were produced in just two states, Iowa and North Carolina. Also illustrating the regionalization, Alabama, Arkansas, and Georgia account for over 30% of U.S. broiler (chicken) production.

Livestock and poultry manure can contain a variety of pathogens. Some are host-adapted and, therefore, not a health risk for humans. Others can produce infection in humans and are thus termed zoonotic. The more common zoonotic pathogens in manure include *Escherichia coli* 0157:H7, *Campylobacter*, *Salmonella*, *Cryptosporidium parvum*, and *Giardia lamblia*. Viruses can also be associated with manure, although less is known about their survival in manure. Survival of microorganisms in manure, soils, and water varies greatly (from days to as much as a year) depending upon the organism and the environmental conditions. Risks from manure-associated pathogens can arise when runoff, spills, or infiltration enable microorganisms to reach surface water or groundwater, or when land-applied manure, or irrigation water impacted by manure, comes into contact with food crops. The level of risk to humans depends upon a number of factors that dictate how readily the microorganisms are transported through the environment and how long they remain infectious, as well as the numbers of microbes and their infectious doses. Most outbreaks of waterborne and foodborne gastrointestinal illness, even those caused by zoonotic pathogens, are attributable to human fecal contamination, although agricultural sources have been implicated in a number of cases. With current surveillance, the degree to which manure-related pathogens may be involved in outbreaks is poorly understood due to difficulties in identifying etiologic agents and sources of contamination, and also because many cases of illness go unreported.

It is estimated that most (60%-80%) livestock and poultry routinely receive antimicrobials. Antimicrobials may be administered to treat and prevent diseases and outbreaks, or at sub-therapeutic levels to promote animal growth and feed efficiency. The U.S. Food and Drug Administration (USFDA) reported that 28.8 million pounds of antimicrobials were sold for animal use in 2009; some estimates suggest this is four times greater than what was used for human health protection during that same year. However, available data are limited and detailed use estimates vary. The overuse and/or misuse of antimicrobials (in general) can facilitate the development and proliferation of antimicrobial resistance, an issue of concern for animal and human health protection. Research indicates that antimicrobial use in livestock and poultry has contributed to the occurrence of antimicrobial-resistant pathogens found in livestock operations and nearby environments. USDA surveys reported that 74% of *Salmonella* and 62% of *Campylobacter* isolates from swine manure were resistant to two or more antimicrobials. Most antimicrobial resistance related to human health is likely the result of overuse and misuse of certain medications in humans. The overlap between livestock and human antimicrobial use is also recognized as an area of concern for human health because the effectiveness of these medications in treating human infections may be compromised. The USFDA banned the use of fluoroquinolones in poultry in 2005 because of human health concerns. The extent to which antimicrobial-resistant human infections are related to the use of antimicrobials in livestock and poultry, is unclear and would benefit from further research.

Hormones are naturally produced by, and in some cases artificially administered to, livestock and poultry. Beef cattle may be treated with hormones to improve meat quality and promote animal growth; dairy cows may be treated to control reproduction and increase milk production. An estimated 720,000 pounds of natural and synthetic hormones were excreted by livestock and poultry in 2000. Research indicates that hormones and their metabolites may be present in environments and surface waters proximal to livestock and poultry operations. While typically detected at low concentrations in water, hormones are biologically active at very low levels and are classified as endocrine disruptors. In aquatic ecosystems, hormones may affect the reproductive biology and fitness of aquatic organisms. Because hormones are excreted by all mammals, including humans, the majority of research has focused on hormone releases from waste water treatment plant discharges. Limited recent research suggests that exposure to hormones from livestock operations and manure may adversely impact the reproductive endocrinology of some fish. More research on the use, occurrence, fate, and transport of natural and synthetic hormones from production facilities and cropland treated with manure is necessary to fully understand their potential impact.

Manure discharges to surface waters can be caused by rain events, spills, storage lagoon and equipment failures, or the improper application of manure, including application to frozen or saturated ground. In some cases, fish mortalities may be caused by oxygen depletion or ammonia toxicity from large loadings of manure. In addition, while cases are limited, nutrients from livestock and poultry manure have been indicated as a cause of harmful algae blooms in surface waters. Harmful algae blooms produce cyanotoxins that may be harmful to animals and aquatic life, as well as to humans when exposed in recreational waters or from drinking water supplies. Proper management and maintenance of lagoons, and minimizing winter land application of manure all help prevent manure discharges to surface waters.

A combination of source water protection, manure management, and water treatment processes can help reduce surface water pollution and remove contaminants from drinking water. While most research has focused on pathogen removal during drinking water treatment, a limited base of recent research has provided some insight into antimicrobial and hormone removal. A stronger understanding of the prevalence and concentrations of antimicrobials and hormones in drinking water, as well as research on which treatment processes best remove these compounds, will help in planning strategies to minimize their consumption and any potential associated health effects.

Good manure management practices, which include the beneficial use of treated manure, linked to sound nutrient management, can help to minimize many problems related to other contaminants. The USDA and their state partners provide technical and financial assistance, as well as conservation practice standards for nutrient and manure management. This report provides a brief introduction to existing programs. The review is not exhaustive, however it provides links to additional information for individuals working in water quality programs.

# Table of Contents

# List of Figures

# List of Tables

# 1. Introduction

This *Literature Review of Contaminants in Livestock and Poultry Manure and Implications for Water Quality* was prepared as part of the United States Environmental Protection Agency's (USEPA) ongoing efforts to better understand the environmental occurrence and potential effects related to contaminants of emerging concern. The report makes no policy or regulatory recommendations; it does identify information gaps that may help define research needs for USEPA and its federal, state and local partners to better understand these issues.

Over the past 60 years the structure of American agriculture has significantly changed. Across all agricultural sectors, farm operations have expanded – farms have gotten larger and fewer in number. The shift from the "family farm" is perhaps most pronounced in the production of livestock and poultry. Since the 1950s, the production of livestock and poultry in the United States (U.S.) has more than doubled, however the number of operations has decreased by 80% (Graham and Nachman 2010). Food animal production has evolved from largely grazing animals and on-farm feed production to fewer and larger operations and increasingly more to concentrated facilities, often with animals raised in confinement (Ribaudo and Gollehon 2006, MacDonald and McBride 2009). This has been done, in part, to meet the demands for meat and animal products from a growing human population in the U.S. and abroad.

The increase in concentration of livestock and poultry also leads to increased concentration of animal manure that must be managed. As production has shifted to much larger, more concentrated operations, livestock and poultry operations have become separated from the land base that produces their feed (Gollehon et al. 2001). Historically, manure was used as fertilizer on the farm to provide nutrients for plant growth on the cropland, pasture or rangeland that, in turn, partly provided the feed for the animals raised on the farm. Manure can also improve soil quality, when managed appropriately as a fertilizer, where the producer considers the right rate, timing, source, and method of application (NRC 1993). However, while livestock manure can be a resource, it can also degrade environmental quality, particularly surface and ground water if not managed appropriately (Kumar et al. 2005). The geographic concentration of livestock and poultry can lead to concentrations of manure that may exceed the needs of the plants and the farmland where it was produced. A report from the U.S. Department of Agriculture's Economic Research Service (USDA ERS) found that more than 60% of manure nitrogen and 70% of manure phosphorus cannot be assimilated by the farmland on which it is generated (Gollehon et al. 2001). Runoff related to manure is considered a primary contributor to widespread nutrient water quality pollution in the U.S., as described in the 2009 "An Urgent Call to Action" report generated by the Nutrient Innovations Task Group (see also Gollehon et al. 2001, Ruddy et al. 2006, Dubrovsky et al. 2010).

While manure's contributions to nutrient water quality impairment is perhaps its most widely recognized impact, manure and livestock management practices may now also be a source of other contaminants (see Table 1-1). Manure often contains pathogens (many of which can be infectious to humans), heavy metals, antimicrobials, and hormones that can enter surface water and ground water through runoff and infiltration potentially impacting aquatic life, recreational waters, and drinking water systems (Gullick et al. 2007, Rogers 2011). The shift towards concentrated livestock production has led to other practices that can contribute contaminants other than nutrients to the environment. To improve animal production efficiency and counteract the greater potential susceptibility of disease in concentrated and confined living conditions, livestock and poultry may be treated with antimicrobials to treat or prevent diseases and infections or treated sub-therapeutically to promote animal growth (McEwen and Fedorka-Cray 2002). Some livestock and poultry also receive steroid hormones to promote animal growth and/or control reproductive cycles (Lee et al. 2007). Pesticides are used to control insect and fungal infestations and parasites as well as other pests. Heavy metals, such as zinc, arsenic, and copper are sometimes added as micronutrients to promote growth.

**Table 1-1. Key pollutants from livestock operations and animal manure.**

| Pollutant | Description of Pollutant | Pathways to the Environment | Potential Impacts |
|---|---|---|---|
| Nitrogen | Organic forms (e.g., urea) and inorganic forms (e.g., ammonium and nitrate) in manure may be assimilated by plants and algae. | • Overland discharge<br>• Leachate into ground water<br>• Atmospheric deposition as ammonia | • Eutrophication and harmful algal blooms (HABs)<br>• Ammonia toxicity to aquatic life<br>• Nitrate linked to methemoglobinemia |
| Phosphorus | As manure ages, phosphorus mineralizes to inorganic phosphate compounds that may be assimilated by plants. | • Overland discharge<br>• Leachate into ground water (water soluble forms) | • Eutrophication and HABs |
| Potassium | Most potassium in manure is in an inorganic form available for plant assimilation; it can also be stored in soil for future plant uptake. | • Overland discharge<br>• Leachate into ground water | • Increased salinity in surface water and ground water |
| Organic Compounds | Carbon-based compounds decomposed by micro-organisms. Creates biochemical oxygen demand because decomposition consumes dissolved oxygen in the water. | • Overland discharge | • Eutrophication and HABs<br>• Dissolved oxygen depletion, and potentially anoxia<br>• Decreased aquatic biodiversity |
| Solids | Includes manure, feed, bedding, hair, feathers, and dead livestock. | • Overland discharge<br>• Atmospheric deposition | • Turbidity<br>• Siltation |
| Salts | Includes cations (sodium, potassium, calcium, magnesium) and anions (chloride, sulfate, bicarbonate, carbonate, nitrate). | • Overland discharge<br>• Leachate into ground water | • Reduction in aquatic life<br>• Increased soil salinity<br>• Increased drinking water treatment costs |
| Trace Elements | Includes feed additives (arsenic, copper, selenium, zinc, cadmium), trace metals (molybdenum, nickel, lead, iron, manganese, aluminum), and pesticide ingredients (boron). | • Overland discharge<br>• Leachate into ground water | • Aquatic toxicity at elevated concentrations |
| Volatile Compounds Including Greenhouse Gases | Includes carbon dioxide, methane, nitrous oxide, hydrogen sulfide, and ammonia gases generated during manure decomposition. | • Inhalation<br>• Atmospheric deposition of ammonia | • Eutrophication<br>• Human health effects<br>• Climate change |
| Pathogens | Includes a range of disease-causing organisms, including bacteria, viruses, protozoa, fungi, prions and helminths. | • Overland discharge<br>• Potential growth in receiving waters | • Animal, human health effects |
| Antimicrobials | Includes antibiotics and vaccines used for therapeutic and growth promotion purposes. | • Overland discharge<br>• Leachate into ground water<br>• Atmospheric deposition | • Facilitates the growth of antimicrobial-resistance<br>• Unknown human health and aquatic life effects |
| Hormones | Includes natural and synthetic hormones used to promote animal growth and control reproductive cycles. | • Overland discharge<br>• Leachate into ground water | • Endocrine disruption in fish<br>• Unknown human health effects |
| Other Pollutants | Includes pesticides, soaps, and disinfectants. | • Overland discharge<br>• Leachate into ground water | • Unknown human health and ecological effects<br>• Potential endocrine disruption in aquatic organisms |

*Adapted from USEPA (2002a) Exhibit 2-2.*

Livestock and poultry operations and related manure management practices account for 18% of all human-caused greenhouse gas emissions (Steinfeld et al. 2006); ruminant livestock and liquid manure handling facilities account for nearly 30% of methane emissions from anthropogenic activities (USEPA 2011a). Besides greenhouse gas emissions, air quality degradation, particularly from concentrated livestock and poultry operations, has been documented, related to releases of toxic as well as odorous substances, particulates, and bioaerosols containing microorganisms and human pathogens (Merchant et al. 2005). Air quality degradation has been related to human health concerns for workers in confined operations and also for neighbors to large facilities (Donham et al. 1995 and 2007, Merchant et al. 2005, Mirabelli et al. 2006).

Recognizing the potential for human and ecological health effects associated with the other contaminants in manure, this report focuses on the growing scientific information related to contaminants of emerging concern – particularly pathogens, antimicrobials, and hormones in manure – and reviews the potential and documented human health and ecological effects associated with these manure contaminants. Many other groups and initiatives are focusing on nutrient water quality issues (i.e., Nutrient Innovation Task Group (NITG) 2009, Dubrovsky et al. 2010), including the relative contributions of animal manure. This report briefly discusses the magnitude of manure generation (which is often highly localized) for perspective on the relationship to these emerging contaminants and their prevalence in the environment, for major livestock types – beef and dairy cattle, swine, poultry and aquaculture. Sections that follow summarize information on pathogens, antimicrobials, and hormones, followed by a review of known or associated impacts related to manure. These sections are followed by a brief review of drinking water treatment methods that can help to deal with contaminants that may be related to manure (and other sources). And the last section of the report provides some direction to other resources and information on manure management. Following good manure management practices which include alternative uses of manure that are both economically and environmentally sustainable, linked to sound nutrient management, can help to minimize many problems related to other contaminants. The USDA NRCS provides technical and financial assistance as well as conservation practice standards for nutrient and manure management.

This report is focused on manure and does not address other waste management issues related to livestock and poultry operations (e.g., disposal of dead animals, spoiled feed). The purpose of this report is to summarize publicly available literature for those involved with watershed protection and management and the linked efforts for source water protection and planning for drinking water systems. As noted in the report, there are very different levels of information available on many of these topics associated with manure. Hence, the report can also help to identify information gaps and guide research needs for the U.S. Environmental Protection Agency (USEPA) and other partners to better understand these issues.

This page intentionally left blank.

# 2. Distribution of Livestock, and Manure Generation and Management

## 2.1. Background

Livestock and poultry production in the U.S. has changed significantly since the 1960's, transitioning towards larger operations separated from the land base that produces their feed (Graham and Nachman 2010). Also, large operations now typically specialize in production of one animal type, often at one stage of its lifecycle (MacDonald and McBride 2009). For example, in swine production, hogs may be transferred from a farrow-to-feeder farm during the initial life stages, to a feeder-to-finish farm and finally to a slaughter plant, rather than being raised at one facility (MacDonald and McBride 2009). The majority of animals are also now raised in confinement where feed is brought to the animal rather than the animals seeking feed in a pasture or on the range (Ribaudo and Gollehon 2006).

Because of the shift in farming practices towards larger animal feeding operations, livestock and poultry production has become more regionalized, and large volumes of manure are oftentimes generated relative to smaller land areas for application (Gollehon et al. 2001). In some areas, the large quantity of manure generated by large operations relative to the small area available for land application magnifies the potential environmental and human health impacts associated with manure runoff and discharges to surface water and ground water.

The mass of manure generated is related to the mass, or size of the animals involved. For example, an average 160-pound human produces approximately two liters of waste per day (feces and urine), whereas an average 1,350-pound lactating dairy cow generates 50 liters of manure (including urine) per day (Rogers 2011). Most animal manure is applied to cropland or grasslands without treatment. Nutrients may be assimilated by the growing plants on cropland and grassland (Graham and Nachman 2010). Through manure storage, handling, and land application, the contaminants associated with manure (i.e., pathogens, antimicrobials, hormones, etc.; see Table 1-1) have the potential to enter the environment (Kumar et al. 2005, Lee et al. 2007, PCIFAP 2008).

> ✓ In 2007, 2.2 billion livestock generated an estimated 1.1 billion tons of manure (as excreted).
>
> ✓ In 1998, USEPA estimated that the livestock manure produced was 13 times greater than all the human sewage produced in the U.S.
>
> ✓ From 1997 to 2007, the number of swine produced in the U.S. increased by 45%, but the number of swine farms decreased by over 30%, resulting in more concentrated manure generation. Over 40% of all swine were produced in just two states: Iowa and North Carolina.
>
> ✓ Cattle (beef, dairy, and other) produce about 80% of all livestock manure in the U.S. – the top 10 producing states produce about 56% of the total.

## 2.2. Cattle, Poultry and Swine

This report uses USDA's 2007 Census of Agriculture livestock and poultry inventory counts to illustrate the distribution of the major animal types (beef and dairy cattle, swine, and poultry) in the U.S. and related manure generation. These tables presented below (and in Appendix 1), summarizing this information by state, are simply to provide perspective on the differences that are apparent around the U.S., and to provide insight on the magnitude of the issues at the state and regional level. These comparisons are made using standard conversion factors developed by the USDA's Natural Resources Conservation Service (NRCS); livestock and poultry counts were converted to animal units (AU), which are a unit of measure based on animal weight

(1 AU = 1,000 pounds live animal weight) (see for example Kellogg et al. 2000, Gollehon et al. 2001). For example, one beef cow or steer equals one AU, whereas it takes 250 layer chickens to equal one AU. The amount of manure generated is directly related to animal weight. Therefore, converting animal counts to AUs allows for the estimation of livestock manure generation and is also a method for standardizing farm operation size across livestock types (Gollehon et al. 2001). (For further information on AU and manure generation calculations, refer to Appendix 1). Several USDA and United States Geological Survey (USGS) reports (i.e., Kellogg et al. 2000, Gollehon et al. 2001, Ruddy et al. 2006) have calculated livestock manure generation using the 1997 USDA Census of Agriculture data. Their estimates, and those presented in this report, are very similar in number, scope, and perspective. (These reports, and this current report, all use the same basic conversion factors noted, but the USDA reports also incorporate more detailed livestock marketing data). The USDA and USGS reports present results at a more detailed scale (i.e., county, watershed, or farm-level manure production), and have been focused on nutrients and nutrient management. Livestock and poultry distribution and manure generation are summarized below (more complete and detailed state-by-state livestock inventories and estimates of manure generation are tabulated in Appendix 1).

In 2007, approximately 2.2 billion cattle, swine, and poultry were produced in the U.S. (USDA 2009a), generating an estimated 1.1 billion tons of manure (manure estimates used here are as excreted, wet-weight). Cattle include beef cattle, dairy cattle, and other cattle and calves (such as breeding stock). Swine include market hogs, which are sent to slaughter after reaching market weight, and breeder hogs, which are used for breeding purposes. Poultry includes chickens as broilers (raised for meat), and as layers (produce eggs), and turkeys. Note that the Census of Agriculture numbers do not account for all the marketing of animals that takes place during a year, and end-of-year 2007 counts were used for analyses. Different than cattle, poultry have a high turnover rate throughout the year. For example, broiler chickens are typically sent to slaughter after five to nine weeks (MacDonald and McBride 2009).

**Table 2-1. Top ten states with the highest beef cattle production and associated manure generation in 2007.**

| National Rank | State | Total Beef Cattle AUs | Percent of Total Beef Cattle AUs* | Total Estimated Tons Manure |
|---|---|---|---|---|
| 1 | TEXAS | 5,259,843 | 16.0% | 60,488,195 |
| 2 | MISSOURI | 2,089,181 | 6.4% | 24,025,582 |
| 3 | OKLAHOMA | 2,063,613 | 6.3% | 23,731,550 |
| 4 | NEBRASKA | 1,889,842 | 5.8% | 21,733,183 |
| 5 | SOUTH DAKOTA | 1,649,492 | 5.0% | 18,969,158 |
| 6 | MONTANA | 1,522,187 | 4.6% | 17,505,151 |
| 7 | KANSAS | 1,516,374 | 4.6% | 17,438,301 |
| 8 | TENNESSEE | 1,179,102 | 3.6% | 13,559,673 |
| 9 | KENTUCKY | 1,166,385 | 3.6% | 13,413,428 |
| 10 | ARKANSAS | 947,765 | 2.9% | 10,899,298 |
| | *Top Ten Subtotal* | *19,283,784* | *59%* | *221,763,516* |
| | **U.S. TOTAL** | **32,834,801** | | **377,600,212** |

*\* Animal units (AUs) represent 1,000 pounds of live animal weight, or one beef cattle per AU (see Kellogg et al. 2000, Gollehon et al. 2001). See Appendix 1 for complete listing of all states. Reference: Inventory data from USDA 2009a.*

The changes in livestock and poultry production – the shift towards fewer, larger, more concentrated production facilities – has resulted in regional and local differences in the distribution of the 2.2 billion animals raised in the U.S. These differences will in turn relate to differences in the issues involved in manure management and the potential for environmental impacts of various contaminants. For example, beef cattle are produced predominantly in the Great Plains and Midwest. According to USDA's 2007 Census of Agriculture, Texas alone accounts for 16% of U.S. beef cattle production with an estimated 60.5 million tons of manure generated – two and a half times greater than the amount generated by the second largest beef cattle producing state (Table 2-1). In contrast, swine are largely produced in Iowa and North Carolina, accounting for 27% and 16%, respectively, of total U.S. production (Table 2-2). Broiler production is predominantly based in the southern and eastern U.S., with Georgia, Arkansas, and Alabama accounting for nearly 30% of U.S. production. An estimated 20.3 million tons of manure from broiler chickens was generated in those three states in 2007 (Table 2-3).

**Table 2-2. Top ten states with the highest total swine (market and breeder hogs) production and associated manure generation in 2007.**

| National Rank | State | Total Swine AUs | Percent of Total Swine AUs* | Total Estimated Tons Manure |
|---|---|---|---|---|
| 1 | IOWA | 2,409,994 | 27.0% | 31,912,337 |
| 2 | NORTH CAROLINA | 1,382,252 | 15.5% | 17,056,820 |
| 3 | MINNESOTA | 999,762 | 11.2% | 12,767,962 |
| 4 | ILLINOIS | 607,844 | 6.8% | 7,289,960 |
| 5 | INDIANA | 486,599 | 5.5% | 6,140,286 |
| 6 | NEBRASKA | 462,548 | 5.2% | 5,543,892 |
| 7 | MISSOURI | 435,930 | 4.9% | 5,252,950 |
| 8 | OKLAHOMA | 367,821 | 4.1% | 4,140,186 |
| 9 | KANSAS | 256,349 | 2.9% | 3,171,100 |
| 10 | OHIO | 243,700 | 2.7% | 3,066,558 |
| | *Top Ten Subtotal* | *7,652,800* | *86%* | *96,342,051* |
| | **U.S. TOTAL** | **8,910,943** | | **111,256,177** |

*\* Animal units (AUs) represent 1,000 pounds of live animal weight (see Kellogg et al. 2000, Gollehon et al. 2001). See Appendix 1 for complete listing of all states. Reference: Inventory data from USDA 2009a.*

Manure management is inherently a local issue, related to the number and type of animals, the land base for application of the manure, the type of operations (i.e., confined feeding operations), and many management factors. Detailed information on all these factors is more difficult to come by, and such estimates are not the purpose or within the scope of this report. (The USDA's Census of Agriculture also does not provide this information (Gollehon et al. 2001)). However, in 2002, a comprehensive review of state livestock production programs was conducted on behalf of USEPA to provide estimates of the number of Animal Feeding Operations (AFOs) and Concentrated Animal Feeding Operations (CAFOs) in each state (Tetra Tech, Inc. 2002). According to that study, the states that had the most AFOs with more than 1,000 AUs were Iowa, North Carolina, Georgia, and California.

**Table 2-3. Top ten states with the highest broiler chicken production and associated manure generation in 2007.**

| National Rank | State | Total Broiler AUs | Percent of Total Broiler AUs* | Total Estimated Tons Manure |
|---|---|---|---|---|
| 1 | GEORGIA | 517,363 | 14.7% | 7,744,926 |
| 2 | ARKANSAS | 444,830 | 12.6% | 6,659,104 |
| 3 | ALABAMA | 391,953 | 11.1% | 5,867,541 |
| 4 | MISSISSIPPI | 330,982 | 9.4% | 4,954,799 |
| 5 | NORTH CAROLINA | 329,498 | 9.4% | 4,932,592 |
| 6 | TEXAS | 260,686 | 7.4% | 3,902,473 |
| 7 | MARYLAND | 143,964 | 4.1% | 2,155,138 |
| 8 | DELAWARE | 112,291 | 3.2% | 1,680,999 |
| 9 | KENTUCKY | 109,399 | 3.1% | 1,637,707 |
| 10 | MISSOURI | 102,537 | 2.9% | 1,534,984 |
| | _Top Ten Subtotal_ | _2,743,505_ | _78%_ | _41,070,264_ |
| | **U.S. TOTAL** | **3,522,083** | | **52,725,576** |

_* Animal units (AUs) represent 1,000 pounds of live animal weight, or 455 broilers per AU (see Kellogg et al. 2000, Gollehon et al. 2001). See Appendix 1 for complete listing of all states. Reference: Inventory data from USDA 2009a._

While manure use and management is a local issue, the state data can also provide some illustrations and valuable perspectives. Table 2-4 summarizes the top ten states related to manure production (this is the sum of the AUs for all livestock, swine, and poultry, and the estimated manure production, as excreted; see Appendix 1). As might be expected, the list is comprised of the major agricultural states, including Texas, Iowa, and California. Texas accounts for about 12% of the AUs and manure produced in the U.S. Total AUs and manure are dominated by beef and dairy numbers because of their body size. Nationally, cattle were responsible for nearly 83% of total livestock manure generation in 2007, followed by swine (10%) and poultry (7%). Refer to Appendix 1 for complete livestock and poultry production and manure generation tables.

As discussed, many of the concerns for environmental impacts of manure generation relate to settings where there is a large mass of manure but a relatively small land base for application of the manure. Even at the state level, these differences can be illustrated. The top livestock states, such as Texas, California, and Iowa (Table 2-4) also have large areas of farm land. Presenting total manure generation on a farmland area basis paints a different picture. Table 2-5 shows the state level estimate for tons of manure generated per farmland acre. Smaller states along the eastern seaboard rise to the top of the list; these states are key poultry and swine producing states but have far more limited farmland than the major farm states. (This tabulation divides the total estimated manure for livestock and poultry by the acreage for "land in farms" from the 2007 Census of Agriculture (USDA 2009a). "Land in farms" is defined by the USDA (2009a) as primarily agricultural land used for grazing, pasture, or crops, but it may also include woodland and wasteland that is not under cultivation or used for grazing or pasture, provided it is on the farm operator's operation. This is an oversimplification at the state level: land in farms is an overestimate of the actual land likely available for application of manure; manure as excreted is likely an overestimate of the mass of manure to be handled, dependent on the management practice. However, it illustrates the differences that are inherent in the distribution of the different types of livestock and poultry settings around the U.S.

**Table 2-4. Top ten livestock and poultry manure producing states in 2007.**

| National Rank | State | Total AUs | Percent of Total U.S. Manure | Total Estimated Tons Manure |
|---|---|---|---|---|
| 1 | TEXAS | 11,109,770 | 11.5% | 128,048,896 |
| 2 | CALIFORNIA | 5,235,439 | 6.2% | 68,496,143 |
| 3 | IOWA | 5,586,515 | 6.1% | 68,360,493 |
| 4 | NEBRASKA | 5,235,899 | 5.3% | 59,100,556 |
| 5 | KANSAS | 4,932,902 | 5.0% | 55,792,510 |
| 6 | OKLAHOMA | 4,571,012 | 4.7% | 52,036,892 |
| 7 | MISSOURI | 4,178,962 | 4.3% | 48,070,611 |
| 8 | WISCONSIN | 3,213,092 | 3.8% | 42,531,594 |
| 9 | MINNESOTA | 3,268,570 | 3.6% | 39,816,914 |
| 10 | SOUTH DAKOTA | 3,179,772 | 3.3% | 36,358,712 |
|  | **U.S. TOTAL** | **92,969,509** |  | **1,113,232,385** |

*\* Data estimated from USDA's 2007 Census of Agriculture livestock counts converted to animal units, following USDA's NRCS methodology. Reference: USDA 2009a.*

**Table 2-5. Top ten states with the highest manure generation in 2007 on a farmland area basis.**

| National Rank | State | Estimated Tons Manure/Acre Farmland\* |
|---|---|---|
| 1 | NORTH CAROLINA | 3.85 |
| 2 | DELAWARE | 3.81 |
| 3 | VERMONT | 3.05 |
| 4 | PENNSYLVANIA | 2.99 |
| 5 | WISCONSIN | 2.80 |
| 6 | CALIFORNIA | 2.70 |
| 7 | NEW YORK | 2.66 |
| 8 | MARYLAND | 2.23 |
| 9 | VIRGINIA | 2.22 |
| 10 | IOWA | 2.22 |

*\* Refer to Appendix 1 for further description on livestock manure generation calculations. Reference: USDA 2009a.*

The way in which livestock and poultry are raised differs by animal type as well as the size of the production facility. Chapter 8 provides further information on manure management programs and strategies. Beef cattle tend to be raised outdoors in pens or corrals, where the manure accumulates and is scraped up along with any bedding materials and soil (in pens), stored in a facility, or stockpiled until it can be land applied on or off-site (USEPA 2009a). In larger, concentrated operations, drainage ditches may flow through beef cattle operations, discharging stormwater, manure, animal feed, bedding materials, and other waste to a nearby collection pond or lagoon (Gullick et al. 2007). Dairy cows may be housed in tie stall barns, free stall barns, or outdoor open lots (USEPA 2009b). Dairy cow manure may be scraped from indoor barns and temporarily stored in a solid stack in steel or concrete tanks, or flushed from barn surfaces and discharged to lagoons (Zhao et al. 2008). Swine are typically housed over slatted floors, allowing manure to be washed down and routinely flushed out of the housing facility (Gullick et al. 2007). Swine manure may be flushed to an underground pit (57% of operations), a lagoon (23% of operations), or another storage area, like a manure pile (20% of operations)

(USDA 2002a). Poultry, including broilers, layers, and turkeys, are almost always raised indoors with manure accumulating and mixing with bedding material (Zhao et al. 2008). Most layers are housed in elevated cages, allowing manure to accumulate below or drop onto a conveyer belt that removes the manure from the building (Gullick et al. 2007). Manure from layers is typically washed from the housing facility to a storage pit (Zhao et al. 2008).

Swine and dairy cow production, in particular, have become increasingly concentrated. Between 1997 and 2007, there was a 33% decrease in the number of swine farms yet a 45% increase in the number of swine processed (USDA 2009a). As shown in Table 2-2, 86% of all U.S. swine production in 2007 occurred in the top ten swine producing states, and the top five states alone account for over two-thirds of U.S. production. From 1997 to 2007 there was a 44% decrease in the number of dairy farms in the U.S., yet the number of dairy cows has remained relatively level, increasing by 1% during that time period (USDA 2009a).

## 2.3. Aquaculture

Aquaculture is a unique component of commercial animal production, very directly related to water resources, and it is also discussed in this report where information is available. The aquaculture sector of U.S. agriculture has been steadily increasing, with a rise in demand for seafood coinciding with declining wild fish and shellfish populations; in providing controlled conditions it may offer production advantages of selective breeding as well as improved disease control (Cole et al. 2009). The USDA's 2005 Census of Aquaculture reported over 4,300 aquaculture farms in the U.S., covering nearly 700,000 acres (USDA 2006). Aquaculture operations may be either freshwater or saltwater, producing an array of aquatic organisms. Aquaculture products include food fish (e.g., catfish, salmon, carp), sport fish (e.g., bass, crappie, walleye), ornamental fish (e.g., goldfish, koi), baitfish (e.g., crawfish, fathead minnows), crustaceans (e.g., crawfish, lobsters, shrimp), mollusks (e.g., mussels, oysters), aquatic plants, and other animals (e.g., alligators, snails, turtles) (USDA 2006). According to the USDA's Aquaculture Census, production in 2005 was situated predominantly in the southern U.S., with Louisiana having the highest total number of freshwater and saltwater operations, as well as the most acres used for aquaculture (USDA 2006). Related to regionalized production and larger but fewer farms, in 2005, the top ten states alone accounted for 95% of the total U.S. aquaculture acreage (see Table 2-6), but less than 50% of the nation's aquaculture farms (refer to Appendix 1 for a complete table).

Catfish production was the dominant commodity in U.S. aquaculture in 2005, with nearly one-third of production occurring in Mississippi (USDA 2006). Trout were the second largest commodity – the majority of which were produced in Idaho (USDA 2006). Catfish are typically raised in ponds, while trout are often reared in flow-through raceways. As defined by the USDA's 2005 Aquaculture Census, flow-through raceways are long, narrow, confined structures in which the water flows into one end and exits the other (USDA 2006). Raceways can be closed systems, in which water flows through a series of ponds prior to discharging into a headwater pond that flows back into the system, or they can be directly linked with a river or stream, using the natural flow to flush water through the system and back into a stream.

Waste produced in aquaculture consists of feces, excess feed, dead fish and other aquatic organisms, nutrients, antibiotics, hormones, pesticides, anesthetics, minerals, vitamins, and pigments (Gullick et al. 2007, Cole et al. 2009). As reviewed by Amirkolaie (2011), up to 15% of feed may be uneaten or spilled, and between 60% and 80% of dietary dry matter may be excreted in intensive aquaculture operations. Aquaculture waste may be managed by removing solids from the water via a settling basin or filtration system, after which the solids may be composted or applied to cropland as fertilizer (Gullick et al. 2007).

**Table 2-6. Top ten aquaculture states in 2005.**

| National Rank | State | Total # of Farms | State | Total Farm Acres |
|:---:|:---|---:|:---|---:|
| 1 | LOUISIANA | 873 | LOUISIANA | 320,415 |
| 2 | MISSISSIPPI | 403 | MISSISSIPPI | 102,898 |
| 3 | FLORIDA | 359 | CONNECTICUT | 62,959 |
| 4 | ALABAMA | 215 | ARKANSAS | 61,135 |
| 5 | ARKANSAS | 211 | MINNESOTA | 41,023 |
| 6 | WASHINGTON | 194 | ALABAMA | 25,351 |
| 7 | NORTH CAROLINA | 186 | WASHINGTON | 13,478 |
| 8 | MASSACHUSETTS | 157 | VIRGINIA | 12,555 |
| 9 | VIRGINIA | 147 | CALIFORNIA | 9,340 |
| 10 | CALIFORNIA | 118 | TEXAS | 7,083 |
| *Top Ten Subtotal* | -- | *2,863* | -- | *656,237* |
| U.S. TOTAL | -- | **4,309** | -- | **690,543** |

*\* See Appendix 1 for complete listing of all states and total aquaculture acreage.*
*Reference: USDA 2006.*

## 2.4. Summary and Discussion

Livestock production in the U.S. is a major industry, representing $154 billion in sales in 2007 – nearly a 55% increase since 1997 (USDA 1999, USDA 2009a). In 2007, 77.6 million cattle AUs (beef and dairy), 8.9 million swine AUs, and 6.4 million poultry AUs generated over 1.1 billion tons of manure (see Appendix 1; inventory data from USDA 2009a). Throughout the various stages of livestock production, considerable amounts of manure and associated contaminants can enter the environment, potentially impacting surface water and ground water, through runoff and discharges. According to the USDA, the shift towards large animal feeding operations and confined operations has resulted in the concentration of wastes and other changing production practices (MacDonald and McBride 2009). Livestock and poultry production has become more concentrated, and larger volumes of manure are generated relative to local land areas where it may be applied; with limited farmland available for manure application, the potential for environmental impacts is of increased concern (Gollehon et al. 2001). For example, despite the fact that dairy cow production remained relatively level between 1997 and 2007, the total number of dairy farms in the U.S. decreased by nearly half during that same ten year time period (USDA 2009a), indicative of the shift towards larger livestock production operations.

The remaining chapters of this report focus on livestock excretion of some key contaminants (e.g., pathogens, antimicrobials, hormones), and their stability in the environment. Livestock manure is a source of pathogens that have the potential to cause infections in humans. Widespread livestock antimicrobial use has been shown to facilitate the growth of antimicrobial-resistant bacteria (WHO 2000), and there is evidence of a linkage between antimicrobial-resistant human infections and foodborne pathogens from animals (Swartz 2002). Hormones excreted by livestock also may contribute to risks to aquatic life, potentially impacting fish reproductive fitness and behavior (Lee et al. 2007, Zhao et al. 2008). Chapter 6 of this report provides a review and analysis of the potential human health and ecological impacts of these emerging contaminants associated with manure.

This page intentionally left blank

# 3. Pathogens in Manure

Manure from livestock and poultry contains a variety of pathogens; some are highly host-adapted and not pathogenic to humans, while others can produce infections in humans (USEPA 2002b). Pathogens that are of animal origin but that can be transmitted to humans are termed "zoonotic" and include prions, viruses, bacteria, protozoa, and helminths (Rogers and Haines 2005). Some may infect one type of livestock, while others may infect several types of animals in addition to humans (Cotruvo et al. 2004). Zoonotic pathogens can have serious public health consequences and garner public attention when major outbreaks occur. Animal agriculture has been implicated as a possible source of contamination in a number of significant outbreaks of human illness (see Section 6.5).

Zoonotic pathogens can be difficult to eradicate from livestock and poultry production facilities because some are endemic to the animal (Rogers and Haines 2005, Sobsey 2006). Furthermore, zoonotic pathogens may have a resistant stage in their life cycle (e.g., a cyst or spore) that enhances their survival in the environment and facilitates transmission to other animals or humans through ingestion of fecal-contaminated water or food. Zoonotic pathogens have the potential for transport to ground water and surface water and may be subsequently ingested through recreation or drinking water (see Section 3.4), with potential implications for human and animal health. They may also contaminate food crops through fecally-contaminated runoff or irrigation water or by contact with soil to which manure has been applied (e.g., Pachepsky et al. 2012, Pachepsky et al. 2011, Rogers and Haines 2005) (see Section 6.5).

This chapter will evaluate manure-associated pathogens that may cause human illness and the various factors contributing to human exposure. Sections 3.1 and 3.2 cover pathogen characteristics, infectious doses, and prevalence by livestock type for important select examples. Section 3.3 briefly discusses the occurrence of pathogens in surface water, ground water, and sediments. Survival of pathogens in various environmental media (manure, soil, sediment, and water) is discussed in Section 3.4, and transport in the environment is discussed in Section 3.5.

## 3.1. Types of Pathogens Found in Livestock

A number of pathogens are associated with fecal matter from livestock and poultry, but only a few pose a known or potential threat to humans, including (USEPA 2004a, Rogers and Haines 2005, Sobsey et al. 2006, Pappas et al. 2008, Bowman 2009):

> **Bacteria:** *Escherichia coli (E. coli)* O157:H7 and other shiga-toxin producing strains, *Salmonella* spp., *Campylobacter jejuni*, *Yersinia enterocolitica*, *Shigella* sp., *Listeria monocytogenes*, *Leptospira* spp., *Aeromonas hydrophila*, *Clostridium perfringens*, *Bacillus anthraxis* (in endemic area) in mortality carcasses
>
> **Parasites:** *Giardia lamblia*, *Cryptosporidium parvum*, *Balantidium coli*, *Toxoplasma gondii*, *Ascaris suum* and *A. lumbricoides*, *Trichuris trichuria*
>
> **Viruses:** *Rotavirus*, hepatitis E virus, influenza A (avian influenza virus), enteroviruses, adenoviruses, caliciviruses (e.g., norovirus)

In addition to pathogens (and often in lieu of pathogens), environmental samples can be tested for microbial indicator organisms, which indicate the possibility of fecal contamination (and thus, the possibility of pathogens). Commonly used indicator organisms include fecal coliforms, *E. coli*, and enterococci (Perdek et al. 2003). *Clostridium perfringens* and coliphages also show promise as indicators because they are present in manure from all animals (e.g., Perdek et al. 2003) (*C. perfringens* is a spore-forming bacterium that is common on raw meat and poultry and is a common cause of foodborne illness (CDC 2011a)). Testing for indicator organisms is more efficient and less expensive than testing for a suite of pathogens associated with livestock and poultry runoff. Indicator organisms have been detected in manure and slurry as well as in runoff (e.g., Thurston-Enriquez et al. 2005, Wilkes et al. 2009). Indicators can, however, have different survival and transport capabilities than pathogens and do not always correlate well with illness or with the pathogens themselves (Perdek et al. 2003). As rapid molecular genetic methods of pathogen detection and enumeration gain wider use, reliance on microbial indicators will lessen. In addition, research is ongoing to better understand the relationships between indicators, pathogens, and other environmental variables such as hydrological conditions and persistence in soils environments (e.g., Wilkes et al. 2009; Rogers et al. 2011).

**Table 3-1. Occurrence, infective doses, and diseases caused by some of the pathogens present in manure and manure slurries from cattle, poultry, and swine.**

| Pathogen | Occurrence (% of positive manure samples)* | | | Infective Doses | Human Diseases and Symptoms |
|---|---|---|---|---|---|
| | Cattle | Poultry | Swine | | |
| **Bacteria** | | | | | |
| *Salmonella* spp. | 0.5 - 18 | 0 - 95 | 7.2 - 100 | 100 - 1,000 cells | Salmonella enteritis, Typhoid Fever, Paratyphoid fever (diarrhea, dysentery, systemic infections that spread from the intestinal tract to other parts of the body, abdominal pain, vomiting, dehydration, septicemia arthritis and other rheumatological syndromes) |
| *E. coli 0157:H7* | 3.3 - 28 | 0 | 0.1 - 70 | 5 -10 cells | Enteric colibacillosis (diarrhea with or without bleeding), abdominal pain, fever, dysentery, renal failure, hemolytic-uremic syndrome , arthritis and other rheumatological syndromes |
| *Campylobacter* spp. | 5 - 38 | 57 - 69 | 14 - 98 | < 500 cells | Campylobacter enteritis (diarrhea, dysentery, abdominal pain, malaise, fever, nausea, vomiting, septicemia, meningitis,, Guillain-Barré syndrome (neuromuscular paralysis), arthritis and other rheumatological syndromes |
| *Yersinia enterocolitica* | - | - | 0 - 65 | 10,000,000 cells | Yersiniosis (Intestinal infection mimicking appendicitis, diarrhea, fever, headache, anorexia, vomiting, pharyngitis, arthritis and other rheumatological syndromes) |
| *Listeria* spp. | 0-100 | 8** | 5.9 - 20 | <10,000 cells | Listeriosis (diarrhea, systemic infections, meningitis headache, stiff neck, confusion, loss of balance convulsions miscarriage or stillbirth) |
| **Protozoa** | | | | | |
| *Cryptosporidium* spp. | 0.6 - 23 | 6 - 27 | 0 - 45 | 10 -1,000 oocysts | Cryptosporidiosis (infection that can be asymptomatic, cause acute but short-lived diarrheal illness, cause chronic diarrheal illness, or be quite severe and cholera-like, with cramping, abdominal pain, weight loss, nausea, vomiting, fever, pneumonia, biliary system obstruction and pain) |
| *Giardia* | 0.2 - 46 | - | 3.3 - 18 | 10-25 cysts | Giardiasis (diarrhea, abdominal cramps, bloating, fatigue, hypothyroidism, lactose intolerance, chronic joint pain) |

*References: Rogers and Haines 2005, Pachepsky et al. 2006, Bowman 2009, USEPA 2010a, Ziemer et al. 2010, and USDA 2007a, 2007b, 2009b, and 2010a. , Ho et al. 2007, Weber et al. 1995, Mohammed et al. 2009.*
*\* Percentage of manure samples testing positive for the pathogen. Range of minimum and maximum percentage as reported in the literature. \*\* Based on a single study.*

Information on the prevalence, illnesses (primarily gastrointestinal), and infectious doses (numbers of organisms required to cause infection) associated with some of the bacterial and protozoan agents are provided in Table 3-1. Occurrence indicates the percentage of manure samples in which the pathogen was detected. The subsections below provide brief descriptions of selected bacterial, protozoan, and viral pathogens as well as summaries of the pathogens associated with each animal type.

### 3.1.1. Bacteria

Below are brief summaries of five zoonotic pathogenic bacteria that can cause serious waterborne or foodborne illness and that are associated with animal manure: *Salmonella, E. coli* O157:H7, *Campylobacter, Yersinia enterocolitica,* and *Listeria monocytogenes.* This list is not comprehensive, but includes some of the organisms that figure prominently in illness and mortality.

#### 3.1.1.1. *Salmonella*

Nontyphoidal Salmonellae, the type of *Salmonella* typically associated with the human infection salmonellosis, are found in the gastrointestinal tracts of cattle, poultry, and swine. (The typhoid agents *Salmonella typhii* and *paratyphi* are specific to humans and are therefore not zoonotic). A higher prevalence of *Salmonella* has been detected in larger chicken, dairy cow, and swine animal feeding operations related to increased herd density and size as well as increased shedding of *Salmonella* (Bowman 2009, USEPA 2010a). *Salmonella* prevalence also varies with animal age and type (Soller et al. 2010). The infectious dose for *Salmonella* is estimated to range from 100 to 1,000 cells (Ziemer et al. 2010), and in 2009, nearly 50,000 cases of salmonellosis were reported in the U.S. (CDC 2011b), although that number does not distinguish between foodborne and waterborne cases.

#### 3.1.1.2. *E. coli* O157:H7

Most strains of *E. coli* bacteria are harmless and live in the intestines of healthy humans and other animals (Rosen 2000). *E. coli* O157:H7, however, is a pathogenic strain of the group enterohemorrhagic *E. coli* (EHEC). This strain is an emerging cause of waterborne and foodborne illness and has been implicated in a number of outbreaks (Table 6-3) (Gerba and Smith 2005). *E .coli* O157:H7 is especially dangerous to young children and the elderly. Similarly to *Salmonella*, a higher prevalence of *E. coli* O157:H7 has been detected in larger dairy cow and swine production operations (Bowman 2009). *E. coli* O157:H7 has been found to be more prevalent in the gastrointestinal system and manure of young calves, lambs, and piglets (Hutchinson 2004, Soller et al. 2010) and appears to colonize cattle for one to two months (Rosen 2000). Prevalence tends to vary by season, increasing during warmer, summer months (Hutchison 2004) and decreasing in colder, winter months (Muirhead et al. 2006). In contrast to *Salmonella*, the infectious dose of *E. coli* O157:H7 is quite low, with estimates of 5 to 10 cells (Ziemer et al. 2010).

#### 3.1.1.3. *Campylobacter*

*Campylobacter jejuni* bacteria are commonly transmitted to humans via contaminated water and food (Perdek et al. 2003) and may co-occur with *E. coli* (AWWA 1999). *Campylobacter* prevalence appears to vary depending on the age of the animal, though conflicting results among reports suggest that other environmental (i.e., animal feeding operation size) and animal-specific factors likely influence prevalence. For example, Hutchison (2004) reported higher prevalence of *Campylobacter* in wastes generated by livestock containing young animals (calves, lambs, or piglets), whereas Soller et al. (2010) and USEPA (2010a) reported increased prevalence in older animals. Estimates for infectious dose in humans are generally < 500 organisms (Table 3-1) (Rosen 2000, Pachepsky et al. 2006, Bowman 2009).

### 3.1.1.4.    *Yersinia enterocolitica*

*Yersinia enterocolitica* causes gastroenteritis and is generally known as a foodborne pathogen (Perdek et al. 2003), although *Yersinia* species are also found in water as well as wild and domestic animals (Rosen 2000). *Yersinia enterocolitica* has been detected in swine feces (Olson 2001). In particular, *Yersinia enterocolitica* O:3 is pathogenic to humans and has been found in the tonsils, oral cavities, intestines, and feces of up to 83% of pigs (Olson 2001); pigs are thus considered a primary reservoir for this pathogen (Rosen 2000). The infectious dose may be in the range of millions of bacteria (Rogers and Haines 2005). *Y. enterocolitica* and other *Y. enterocolitica*-like organisms have been isolated from feces of pigs, cattle, and other animals (Brewer and Corbel 1983).

### 3.1.1.5.    *Listeria monocytogenes*

*Listeria monocytogenes* causes severe illness, including diarrhea and meningitis. This bacterium is resistant to adverse environmental conditions (i.e., heating, freezing, and drying). Pathogenic strains are found in ruminants in which they can cause disease (Bowman, 2009). *Listeria monocytogenes* is also found in poultry (Chemaly et al. 2008) as well as sheep, pigs, and other animals (Weber et al. 1995). Levels of *Listeria* spp. can vary by season; Hutchinson (2004) reports that it is more likely to be isolated during March to June (Hutchinson 2004). Husu et al. (2010) reported that prevalence in fecal samples is higher during the indoor season than when the animals are at pasture. According to the USFDA (2012a), the infectious dose for humans may vary widely and depends upon a number of factors, including the strain, susceptibility of the host, and the matrix in which it is ingested. It has been reported to be <10,000 (Table 3-1), but USFDA (2012a) notes that for susceptible individuals consuming raw or inadequately pasteurized milk, it may be as low as 1,000 cells.

## 3.1.2.  Parasites

Three selected types of illness-causing parasites that may be present in manure, *Cryptosporidium parvum*, *Giardia lamblia*, and helminthes (worms) are briefly discussed below. *Cryptosporidium* and *Giardia* cause gastrointestinal illness; infection with helminthes can cause problems that include pneumonia, cysts, or intestinal infections.

### 3.1.2.1.    *Cryptosporidium*

*Cryptosporidium parvum* is a protozoan parasite that can cause cryptosporidiosis, or gastric and diarrheal illness, in humans (Table 3-1) (Rose 1997). Cryptosporidiosis can be contracted through ingestion of small, hardy oocysts from fecally contaminated drinking water supplies, food, recreational waters, pools, and direct contact with animals (Perdek et al. 2003). There is currently no treatment for Cryptosporidiosis, and it can lead to fatality in vulnerable populations such as the immunocompromised. *Cryptosporidium parvum* is shed primarily by relatively young animals (Rosen 2000, Bowman 2009), and upper age estimates for shedding range from 30 days (Rosen 2000) to six months (Atwill 1995). Prevalence is greater during the summer months (Garber et al. 1994, Scott et al. 1994). Cattle can shed substantial quantities of oocysts; estimates include 10 million (Rosen 2000) to more than one billion oocysts per gram of manure (USEPA 2004a), which is orders of magnitude higher than the infectious dose (Table 3-1) (Bradford and Schijven 2002, Pachepsky et al. 2006).

### 3.1.2.2.    *Giardia*

*Giardia lamblia* is the most common cause of protozoan infection in humans (Perdek et al. 2003), causing a gastrointestinal illness known as Giardiasis. Giardiasis can be treated with drugs, and it is not considered a fatal illness. *Giardia lamblia* forms a durable egg-like cell called a cyst through which infection is transmitted, typically via ingestion of fecal-contaminated water (Ziemer et al. 2010). *Giardia* may be present in cattle as young as five days old, up to adults, although prevalence peaks when the calves are young. Prevalence has been reported to range from less than 14% to 100% in calves less than six months old (Rosen 2000, Soller et al. 2010). As with *Cryptosporidium*, the infectious dose for *Giardia* is low (10 to 25 cysts) (Pachepsky et al., 2006), and *Giardia* cysts can be shed in large numbers. According to one study, concentrations of *Giardia* cysts can be over 1,000 cysts/g in swine lagoon wastewater (Ziemer et al. 2010).

### 3.1.2.3.    Helminthes

Helminthes are worms that may be parasitic in plants and animals or may be free-living (NRCS/USDA, 2012). Parasitic worms of concern include Platyhelminthes (flatworms) and Nematoda (roundworms). Some (e.g., most flatworms) have complex lifecycles that require several hosts (Rogers and Haines 2005). The most common parasite in humans is *Ascaris lumbricoides*, a large parasitic roundworm for which humans are the definitive host (NRCS/USDA/2012, Ziemer et al. 2010). Important helminthes that infect livestock include *Ascaris suum* and *Trichuris suis* (cattle and pigs) (Bowman 2009). *Ascaris suum* is associated with swine in particular (Ziemer et al. 2010); its eggs are hardy and can survive in soil and feces for years (Olsen 2001). Illnesses caused by *Ascaris* sp. include pneumonia when the worms invade the lungs or intestinal infection (NRCS/USDA 2012). Infection of humans with zoonotic helminthes generally occurs via consumption of raw or undercooked meat rather than through exposure to feces (Ziemer et al. 2010); these organisms are not discussed further in this chapter.

## 3.1.3. Viruses

A number of viruses, including prevalent enteric viruses that cause gastroenteritis, are present in livestock and poultry and have zoonotic potential. Below are brief descriptions of three common viruses: rotavirus, norovirus, and hepatitis E virus.

### 3.1.3.1.    Rotavirus

Rotavirus is an enteric virus that causes millions of cases of diarrhea in the U.S., primarily in infants and children less than two years of age (Perdek et al. 2003). It has been found in swine, cattle, lambs, and other animals (Cook et al. 2004). There is evidence for zoonotic transmission in that serotypes and genotypes of animal strains have been found in humans, and there is evidence for reassortment (mixing) of genetic material between human and animal rotaviruses (Laird et al. 2003, Cook et al. 2004, Ziemer et al. 2010). The estimated infectious dose for rotavirus is low (10 to 100 virus particles) (Grieg and Todd 2010).

### 3.1.3.2.    Norovirus

Noroviruses are enteric viruses that cause diarrhea in humans as well as livestock in swine and cattle. They are a leading cause of non-bacterial gastroenteritis, estimated to cause more than 90% of outbreaks worldwide (Wang et al. 2006). Swine are believed to serve as an important reservoir for human norovirus, which is closely related to porcine norovirus. Also, there may be reassortment between human and porcine strains (Mattison et al. 2007). A study by Wang et al. (2006) found that noroviruses are found only in finisher hogs, (those ready for slaughter), with a prevalence of 20%. The infectious dose is estimated at 10 to 100 virus particles (Moe et al. 1999).

### 3.1.3.3.    Hepatitis E

Hepatitis E virus (HEV) causes liver inflammation. Humans are the primary reservoir, but swine are also an important reservoir (Perdek et al. 2003, Kasorndorkbua et al. 2005). According to one study, up to 100% of swine tested seropositive for HEV in commercial herds in the Midwestern U.S. (Meng et al. 1997). Another study identified HEV ribonucleic acid (RNA) in about 23% of hogs (Fernández-Barredo et al. 2006). Swine shed the virus for three to four weeks, primarily weaners (hogs being weaned from nursing) and hogs in their first month of feeding (Kasorndorkbua et al. 2005). Swine and human HEV are closely related (Meng et al. 1997). Researchers have noted cross-species infections of human and swine HEV (e.g., Ziemer et al. 2010). The infectious dose is not known (PHAC 2010), nor is its survival in manure known (Ziemer et al. 2010).

## 3.2. Pathogens by Livestock Type

Several of the major zoonotic pathogens, including those described in the previous section, are associated with more than one type of livestock, although the health risks that they pose may vary depending upon the species and prevalence. The following subsections briefly summarize which pathogens associated with cattle, swine, and poultry may cause illness in humans.

### 3.2.1. Cattle

Beef and dairy cattle are carriers of several zoonotic pathogens including *E. coli* O157:H7, *Cryptosporidium parvum*, *Giardia lamblia*, *Campylobacter*, *Leptospira*, various enteroviruses, norovirus, *Listeria monocytogenes*, and *Salmonella* (Cotruvo et al. 2004, Bowman 2009) (Table 3-1). The prevalence of some pathogens has been found to be greater in larger herds (e.g., Bowman 2009, USEPA 2010a; subsections 3.1.1 and 3.2.1). Cattle are an important reservoir of *E. coli* O157:H7, and any herd may contain asymptomatic animals. Estimates of *E. coli* O157:H7 prevalence vary widely. According to a study published for the World Health Organization (WHO), an estimated 30% to 80% of cattle carry *E. coli* O157:H7 (Cotruvo et al. 2004). In contrast, a study of cattle in 13 U.S. states showed that less than 2% of cattle tested positive for the organism (Dargatz 1996). Other estimates range from

> ### *E. Coli* O157:H7 in Cattle
>
> *E. coli* is found frequently among cattle operations. A 1997 survey of 100 feedlots in the U.S. found *E. coli* O157:H7 in 63% of the feedlots tested. However, only 1.8% of manure samples tested positive at these feedlots. Another study found that as many as 28% of beef cattle were shedding *E. coli.* O157:H7, and more than 43% of carcasses tested positive for the bacterium (References: Hancock et al. 1997, Bowman 2009).

about 3% to 28% (Table 3-1; see text box). Cattle are also considered to be a significant source of potential human infection with *Giardia lamblia* (Bowman 2009) and *Cryptosporidium parvum* (Table 3-1).

### 3.2.2. Swine

Swine are hosts to a large number of pathogens including *Campylobacter*, *Yersinia enterocolitica*, *Giardia*, *Salmonella*, *Cryptosporidium*, *E. coli* O157:H7, *Leptospira*, *Balantidium coli*, *Listeria*, and viruses (rotavirus, norovirus, HEV) (Perdek et al. 2003, Rogers and Haines 2005, Mattison et al. 2007, Ziemer et al. 2010, USEPA 2010a). A U.S. survey found that about 80% of pigs older than three months test positive for HEV (Bowman 2009). Swine urine is a potentially important source of *Leptospira*, which has been implicated in waterborne infections (Bowman 2009). Swine *Cryptosporidia* present a lower risk to humans because the species they carry are specifically adapted to swine as a host (USEPA 2010a). These pathogens may be transmitted to humans either through direct contact with swine waste (e.g., workers at an animal feeding operation) or indirectly through the environment (e.g., swimming in manure-contaminated water or consuming contaminated drinking water).

### 3.2.3. Poultry

*Salmonella* and *Campylobacter jejuni* are highly prevalent among poultry in the U.S. (USEPA 2010a), and the serotypes are similar to those implicated in human infections (Ziemer et al. 2010; Rogers and Haines 2005). *Campylobacter butzleri*, now *Arcobacter butzleri*, has also been isolated in poultry (Houf et al. 2003). Chickens do not pose a risk for humans with respect to *Cryptosporidium* and *Giardia;* the *Cryptosporidium* species that infect chickens are a low risk to humans, and chickens do not appear to carry *Giardia* (USEPA 2010a).

---

#### *Campylobacter* in Poultry

*Campylobacter* is found in the intestines of both wild and domestic animals, especially poultry. Flocks may approach 100% infection rates in poultry facilities. *Campylobacter* is commonly (>50%) found in chicken manure and is also associated with swine and, to a lesser degree, cattle manure. The pathogen is typically transmitted via contaminated water and food. *Campylobacter* may co-occur with *E. coli.* (References: AWWA 1999, Cox et al. 2002, Perdek et al. 2003, USEPA 2010a).

---

## 3.3. Occurrence of Pathogens in Water Resources

In the USEPA's 2004 National Water Quality Inventory (USEPA 2009c), microbial contamination was a leading cause of impairment in rivers and streams, with agriculture identified as an important contamination source. Microbial constituents may reach surface water bodies via wet weather flows from animal feeding operations or areas where manure has been land applied or when lagoons are breached. A number of studies have specifically documented effects from pathogens and indicator organisms (see Section 3.1). For example, fecal coliforms and *Streptococcus,* both indicators, have been found in agricultural runoff (Simon and Makarewicz 2009), through which these microorganisms may reach surface water bodies, sometimes contributing to exceedances of water quality standards and possibly to exceedances of permit limits (Baxter-Potter and Gilliland 1988, USEPA 2002b). Work by Kemp et al. (2005) documented *Campylobacter* in surface water due to runoff from dairy farming. In grazing areas, free access of cattle to streams allows manure to reach the water and has been associated with elevated stream bacterial concentrations, with up to 36-fold increases in *E. coli* reported in stream water samples compared to upstream levels (Schumacher 2003, Vidon et al. 2008, Wilkes et al. 2009). Among the protozoa, *Cryptosporidium* oocysts may be carried in runoff, especially after rain events, and *Giardia* cysts have been detected in surface waters as well as ground water (Cotruvo et al. 2004). A study of *Giardia* and *Cryptosporidium* in 66 surface water drinking water sources revealed *Giardia* cysts in 81% of raw water samples and *Cryptosporidium* oocysts in 87% of raw water samples (LeChevallier et al. 1991). Although in general, contamination of water bodies from viruses in manure is less well understood, some authors (e.g., Payment 1989, Rosen 2000, Ziemer et al. 2010) have noted that runoff or waste from lagoons can supply viruses to water bodies (Payment 1989, Rosen 2000, Ziemer et al. 2010). Microbial populations are also found in bottom sediments. They can be present in higher concentrations than in the overlying water column because of the tendency of microbes to associate with particles that settle and because of their improved survival in sediments (see subsection 3.4.2 on factors influencing pathogen survival) (van Donsel and Geldreich 1971, Davies-Colley et al. 2004). *E. coli* and fecal coliform concentrations in sediments have been reported as high as $10^5$ colony forming units per 100 mL (Crabill et al. 1999). When resuspension occurs due to rainstorms or dredging, microorganisms can be released from sediments to the water column (Kim et al. 2010). Spikes in waterborne fecal indicator bacteria have been observed after rainfall (Cho et al. 2010).

Although soil cover and the unsaturated zone provide protection to ground water with respect to pathogen contamination (see subsection 3.5.2), microorganisms can reach ground water. When they do, they may travel downgradient, with the rate of travel depending upon the geologic and hydrogeologic properties of the aquifer. Enteric viruses have been observed to be transported via ground water (Rogers and Haines 2005), and a nationwide survey of drinking water wells revealed enteroviruses in 15% of samples (Abbaszadegan et al. 2003). Bacteria and *Cryptosporidium* oocysts are also believed to have the potential to be transported in ground water; one study documented *E. coli* contamination of ground water downgradient from an unlined cattle manure lagoon (Withers et al. 1998). Ground water in karst areas is particularly vulnerable to contamination because of the channelized nature of the rock, which allows rapid flow and may transport pathogens greater distances. While shallow unconfined aquifers are most vulnerable to contamination, deep, confined aquifers may also be vulnerable to pathogen contamination where there are fractures in the confining layer or from transport along poorly cemented wells (Borchardt et al. 2007).

**Table 3-2. Survival of selected bacterial and parasitic pathogens found in manure, soil, and water.**

| Pathogen | Survival (days)* | | |
|---|---|---|---|
| | **Soil** | **Water** | **Manure** |
| Bacteria | | | |
| *Salmonella* spp. | 16 - 196 | 35 to >186 | 20 to 250 |
| *E. coli 0157:H7* | 2 to >300 | 35 to >300 | 50 to >300 |
| *Campylobacter* sp. | 7 to 56 | 2 to >60 | 1 to 56 |
| *Yersinia enterocolitica* | 10 to >365 | 6 to 448 | 10 to >365 |
| *Listeria* sp. | <120 | 7 to >60 | >240 |
| Protozoa | | | |
| *Cryptosporidium* spp. | 28 to >365 | 70 to >450 | 28 to >400 |
| *Giardia* | < 1 to 28 | < 1 to 77 | < 1 to 77 |

*The range shows the shortest and the longest survival time the organisms can survive at different temperatures for all types of manure (cattle, swine and poultry) and water (surface, ground, and drinking water). References: Rogers and Haines 2005, and Bowman 2009.*

## 3.4. Survival of Pathogens in the Environment

The potential adverse impacts on humans from zoonotic pathogens is directly related to the organisms' survival in various environmental media such as manure, soil, sediments, surface water, and ground water (Cotruvo et al. 2004). Survival of zoonotic pathogens in animal manure and in the environment can range from days to years (Ziemer et al. 2010) depending upon the characteristics of the pathogen and the environmental conditions (Rogers and Haines 2005). The survival capabilities of *Cryptosporidium* oocysts deserve particular mention because of their long survival times in the environment (Ziemer et al. 2010), their resistance to conventional drinking water disinfection processes (chlorine and chlorine dioxide; see Chapter 7) (Edzwald 2010), and the lack of any treatment for human infection. *Cryptosporidium* oocysts can remain viable in a range of environmental settings and can persist in damp conditions for months (Brookes et al. 2004, Ziemer et al. 2010).

The persistence of pathogens in environmental media depends on environmental conditions and the survival characteristics of the microbes present. The factors influencing pathogen survival include temperature, ultraviolet (UV) radiation, moisture, pH, nutrient availability, ammonia concentration in the medium, predation, and competition for nutrients (Rogers and Haines 2005). The sections below include a brief overview of the factors that affect the survival of pathogens in manure, soil, sediments, and water, providing examples relevant to bacteria, protozoa, and viruses.

## 3.4.1. Manure

Manure can provide a favorable environment for pathogen survival and even re-growth due to the availability of nutrients as well as protection from UV radiation, desiccation, and temperature extremes (Rogers and Haines 2005). Conversely, several factors promote die-off in manure, including predation, competition, and the concentration of inorganic ammonia (Rogers and Haines 2005). Temperature in particular is a critical factor in pathogen survival, with cooler temperatures generally enabling longer survival times. Bacterial pathogens such as *Salmonella* and *E. coli* O157:H7 can survive for several months in manure when environmental conditions are favorable (low temperatures, good moisture level) (Rogers and Haines 2005). Increased temperatures, on the other hand, hasten die-off. The extent of this effect varies by organism, but survival in manure generally drops markedly at temperatures exceeding 20 to 30°C compared with survival at cool temperatures (1 to 9°C) (Rogers and Haines 2005). This dependence of survival times on temperature results in seasonal trends; for example, a study of *Salmonella typhimurium* in swine slurry showed survival times of 26 days during summer and 85 days during winter (Venglovsky et al. 2009). As described further in Chapter 8, microorganisms can be inactivated when using certain manure management practices, such as composting, which produces elevated temperature (Olson 2001, Schumacher et al. 2003).

The effects of freezing on pathogen survival vary by organism. Viruses can maintain infectiousness after freezing (Ziemer et al. 2010). *Cryptosporidium* oocysts have been shown to survive freezing in manure and soil for more than three months to one year, but *Giardia* cysts are inactivated (Olson 2001, Rogers and Haines 2005). *Salmonella* is also not inactivated by freezing (Olson 2001). However, the stress of repeated freeze-thaw cycles does generally reduce microbial survival (Rosen 2000).

Compared to bacteria and protozoa, less research has been conducted on the survival of viruses in manure. The available literature, however, suggests that viruses may survive longer than bacteria (Rogers and Haines 2005). For example, extended manure storage (two years) may be required to achieve a 4-log (10,000 fold) reduction in the concentrations of some viruses such as rotavirus (Pesaro et al. 1995). More research is needed on virus survival in manure given the potential for viruses to enter into soil when manure is spread on land and there is a possibility of transport to water and drinking water sources via runoff.

## 3.4.2. Soils

In soils, pathogen survival is influenced by temperature, moisture content, pH, predation, nutrient availability, competition with native soil microorganisms, and organic matter content (Rosen 2000, Unc and Goss 2004). Aside from temperature, moisture exerts an important control, with increased moisture promoting survival (Reddy et al. 1981, Unc and Goss 2003, Venglovsky et al. 2009). Fecal coliform bacteria survive longer in organic soils than in mineral soils, possibly due to the greater capacity of organic soils to hold water (Unc and Goss 2003). Desiccation decreases the survival of *Cryptosporidium*, *Giardia*, fecal bacteria such as *Campylobacter* (Olson 2001, Rogers and Haines, 2005, Bowman 2009), and viruses (Bosch et al. 2006). Predation by native soil organisms can contribute to pathogen removal and has been identified as one of several biological factors in pathogen inactivation that merit further study (Bosch et al. 2006, Rogers and Haines 2005). For viruses, survival in soils has been found to be increased by adsorption to soil as well as decreased soil pH; the pH effect is likely due to greater adsorption of viruses to particles at lower pH (Hurst et al. 1980). For bacteria, however, low pH reduces survival (Unc and Goss 2004).

Exposure to UV light from direct sunlight, such as during land application, can contribute to microbial die-off and is discussed further below. In manure and in soil, microorganisms will associate with particulates, where they are protected from sunlight within the soil profile (e.g., Thurston-Enriquez 2005), especially if manure is worked into soil during application. At the soil surface, however, microbes will be vulnerable to inactivation due to sunlight as well as desiccation (Tyrrel and Quinton 2003).

### 3.4.3. Sediments

Bottom sediments in manure lagoons or natural waters can serve as a very effective reservoir for pathogens because the sediment environment provides moisture, soluble organic matter, and nutrients as well as protection from UV light, desiccation, and predation by protozoa (Rogers and Haines 2005, Cho et al. 2010, Kim et al. 2010). Microorganisms can survive in this environment for long periods of time; fecal bacteria have been shown to survive in sediments from weeks to months (Schumacher et al. 2003, Cho et al. 2010).

### 3.4.4. Water Resources

Pathogen survival in water depends upon a variety of factors including water quality (e.g., turbidity, dissolved oxygen, pH, organic matter content) and environmental conditions (i.e., temperature, predation by zooplankton). Survival times for *Giardia* and *Cryptosporidium* can be quite long (Ziemer et al. 2010); *Cryptosporidium* oocysts can survive from months to more than a year in cold water (5°C) (Ziemer et al. 2010; Olson 2001, Cotruvo et al. 2004, Rogers and Haines 2005). Giardia cysts survive less than 14 days at 25°C but could survive up to 77 days at 4 to 8°C (Ziemer 2010). Enteric viruses, such as the hepatitis E virus and hepatitis A virus tend to be stable in water, especially in colder temperatures (Cotruvo et al. 2004).

Some bacteria (e.g., *Campylobacter* and *E. coli*) can enter a viable but non-culturable state, in which the bacteria's metabolism slows and it cannot be grown in culture media, but it retains infectiousness (Perdek et al. 2003). The viable but non-culturable state can be brought about by low temperatures and stress from starvation, but the cells will reactivate under favorable conditions (e.g., increased temperature). This state has implications for monitoring and may cause contamination to be missed during sampling if culture methods are used for analysis.

As with pathogen survival in manure and soil, exposure to UV light is a key factor in bacterial, viral, and protozoan die-off in surface waters (Rosen 2000, Cotruvo et al. 2004, Fong and Lipp 2005). For example, UV light can cause a reduction of up to four orders of magnitude in the viability of *Cryptosporidium* (Bowman 2009). Ultraviolet light has also been demonstrated to be effective against human enteric viruses and bacteriophages (Kapuscinski and Mitchell 1983, Fujioka and Yoneyam 2002, Battigelli et al. 1993). Greater turbidity of the water, however, affords microorganisms some protection from UV light, and an aquifer environment also protects pathogens against UV exposure and facilitates their survival in ground water.

## 3.5. Transport of Pathogens in the Environment

Pathogens and indicator organisms associated with manure can be transported to surface water and ground water through runoff, discharges, infiltration, and atmospheric deposition (Jawson et al. 1982, USEPA 2002b, Soupir and Mostaghimi 2011). Lagoon spills and flooding of constructed treatment wetlands during severe rainstorms or lagoon leaks and equipment failures during dry weather may also release waste and associated pathogens into the environment (Marks 2001, USEPA 2002b, Rogers and Haines 2005). Tile drainage may also provide a route for microbes in ground water to reach surface waters (Rogers and Haines 2005). The sections below briefly discuss considerations related to transport in runoff, soil infiltration, and transport in ground water.

### 3.5.1. Runoff and Transport to Surface Water

A key mechanism of pathogen transport to surface waters is via runoff (overland flow from rain or snowmelt, or releases from manure pond leaks/overflows). During a rain event, for example, the partitioning of flow between surface runoff and infiltration through the soil depends upon a number of factors. Storm intensity and duration, soil hydraulic characteristics (e.g., permeability, antecedent moisture and temperature), land slope, and soil cover have all been shown to influence runoff and therefore pathogen transport (Rosen 2000, USEPA 2002b). If rainfall intensity exceeds the capacity of the soil to infiltrate water, overland flow occurs, and microorganisms can be carried rapidly in surface runoff (Tyrrel and Quinton 2003, Unc and Goss 2003). Clay-rich soils also tend to promote surface runoff due to their low permeability. Additionally, bare soil with heavy animal traffic can contribute substantial pathogen loads to runoff through erosion of pathogen-laden soil particles (Rosen 2000).

To be available for transport in runoff, pathogens are released from the manure. Most pathogens do remain associated with the fecal deposit during rain events (NRCS/USDA 2012). The amount of pathogens that are released from manure depends upon a number of factors related to the manure itself and the method of application. Important factors include the loading of pathogens in the manure, the pathogen types and survival characteristics, and the age and source of the manure. Aging can greatly reduce the amount of microorganisms that leach out of the manure, due at least in part to declines in the fecal loads in the manure with time and environmental exposure (NRCS/USDA 2012).

The form of manure (solid versus liquid) may affect how easily pathogens reach waterways (e.g., Thurston-Enriquez et al. 2005), with liquid application permitting ready transport via runoff. Also, the amount applied and the style and timing of application will have effects. If manure is applied to frozen ground or immediately before or after a rain event, there will be a greater chance for pathogen transport in runoff. There is uncertainty and limited information, however, regarding whether the method of application (surface application vs. injection) affects runoff quality. Injection may limit runoff from the surface, but UV radiation, heat, and desiccation on the surface would promote die-off. Tyrrel and Quinton (2003) note that some studies have shown no difference in water quality but that their own unpublished data for small scale rain simulation events showed greater (10-fold) fecal coliform transport if waste is surface-applied.

Once pathogens and indicator organisms reach rivers and streams, their transport will be governed by a number of factors including channel morphology, streambed composition, and turbulence and flow regimes (NRCS/USDA 2012). Transport of up to 21 kilometers has been reported for bacteria that were experimentally added to a stream. Microorganisms can be transported either as free organisms (Soupir and Mostaghimi 2011) or associated with soil or manure particles (USEPA 2002b, Pachepsky et al. 2006, Bowman 2009), with free cells in suspension having the potential to travel farther because their small size minimizes settling (Tyrrel and Quinton 2003). Free-living organisms may be added to the streambed sediments when water infiltrates into the streambed (NRCS/USDA 2012).

The amounts of pathogens that become associated with particulates in runoff and surface waters will vary by organism, source, and the particulates available. Studies of stormwater as well as stream and estuarine settings have reported 15% to 35% of bacteria to be associated with particles (Characklis et al. 2005, Cizek et al. 2008, Suter et al. 2011). Also, large fractions of *Giardia* and *Cryptosporidium* (60% and 40%, respectively) have been found to be bound to sediment in streams (Cizek et al. 2008). Microorganisms attached to larger soil particles may settle, especially in quiescent waters, contributing to pathogen loads in bottoms sediments (Rogers and Haines 2005). Microorganisms associated with colloids (very small particles that do not settle) will continue to be transported downstream.

### 3.5.2. Transport through Soil to Ground Water

Transport through the soil profile and in ground water involves an extremely complex interplay of physical and chemical processes that depend upon the size and surface properties of the microorganism; the composition, mineral surface properties, and texture of the soil or aquifer material; the composition of the aqueous medium; and the hydraulic conditions (e.g., saturated vs. unsaturated flow). The following subsections briefly describe some of the features controlling microbial transport and retention.

#### *3.5.2.1.    Physical Processes (Filtration and Flow through Soil)*

Soil generally provides some degree of protection to ground water resources from pathogens by retaining them through physical processes (straining/filtering) and/or through adsorption, particularly in the upper layers of the soil (see subsection 3.5.2.2) (Bicudo and Goval 2003). Fine-grained soils, such as those with greater silt and clay, are most effective at filtering larger bacteria and protozoa (Rosen 2000, Jamieson et al. 2002). Because of their small size, viruses are less likely to be retained in the soil by filtration than bacteria or protozoa (Rosen 2000, USEPA 2004a), although they may be removed by adsorption (see subsection 3.5.2.2). Their small size also renders viruses relatively mobile in ground water (USEPA 2004a).

During heavy rainfall, transport through the soil may be rapid if there is enough water to fill the pore spaces, and microbes may reach the water table more quickly than during lighter rainfall (Unc and Goss 2003, Rosen 2000, USEPA 2004a). Preferential transport may occur through macropores, wormholes, and root channels (Jamieson et al. 2002, USEPA 2004a), bypassing the filtering effect of the soil matrix (Rosen 2000). Wormholes and root channels can be reduced by conventional tillage, but they are not disturbed by conservation tillage or in pasturelands (Bowman 2009). Conditions especially conducive to microbial contamination of ground water include a combination of recent manure application on land with coarse, sandy soil or soil with macropores and a shallow water table (USEPA 2004a, Bowman 2009). Once in ground water, pathogen transport may be particularly rapid in fractured rocks or karst areas because of large channels in the rock.

#### *3.5.2.2.    Retention by Adsorption in Soil and Aquifers*

Adsorption/desorption interactions are extremely important in governing the mobility of microbes. For example, viruses may be removed by adsorption in the first few inches of soil during infiltration, although rainfall can later cause desorption of viruses from the soil, allowing for continued transport and continued contamination (Landry et al. 1979, Goyal and Gerba 1979). Parasites may also be retained. In an experimental study with intact soil cores, *Cryptosporidium parvum* oocysts were mostly retained in the soil within the upper 0.75 inch of soil (Mawdsley et al. 1996), although the authors note that the study was done using purified oocysts, which may not be representative of oocysts in the environment. A number of studies have focused on understanding bacterial sorption to soils and aquifer sediments, with soil and ground water chemistry both playing important roles (e.g. Hendricks et al. 1979, Scholl and Harvey 1992, Banks et al. 2003).

The soil and aquifer characteristics that promote microbial adsorption are: a high clay content, high iron oxyhydroxide and aluminum oxide content, high organic matter, and pH below 7 (e.g., Goyal and Gerba 1979, Rosen 2000). Bacteria tend to adsorb well to ferric oxyhydroxide coatings on clay minerals or quartz through electrostatic attraction (Mills et al. 1994). Organic carbon in the soil contributes to retention of viruses and bacteria due to hydrophobic partitioning (e.g., Rogers and Haines 2005). Furthermore, manure application changes soil pH and adds salts as well as soluble and insoluble organic compounds, altering properties of both the soil and microbes and potentially affecting retention of microbes by the soil (Unc and Goss 2004).

Soil water or ground water characteristics that affect adsorption include pH, ionic strength, divalent cation concentrations, and dissolved organic carbon. Adsorption of viruses to soil particles is enhanced by low pH or increased ionic strength of the water (Rogers and Haines 2005). For bacteria, an increase in ionic strength, particularly due to high divalent cation concentrations, has been shown to increase retention in a sandy medium (e.g., Mills et al. 1994). Dissolved organic matter, on the other hand, has been found to hinder virus adsorption (e.g., Goyal and Gerba 1979, Lance and Gerba 1984). If application of liquid manure or leaching of solid manure by rainfall changes the ionic strength and/or organic carbon content of the soil water or ground water, the capability of the soil or aquifer system to retain microorganisms may change.

## 3.6. Summary and Discussion

Livestock and poultry manure can carry an array of zoonotic pathogens, which can be transported to recreational and drinking water resources. The most common pathogens of concern are *E. coli* 0157:H7, *Campylobacter*, *Salmonella*, *Cryptosporidium parvum*, and *Giardia lamblia*. Other zoonotic organisms include *Listeria* and *Yersinia*, and several viruses may have zoonotic potential (see text box). Infectious doses vary widely among pathogens, and some doses are very low, especially those for *E. coli* 0157:H7 (5 to 10 cells) and the protozoa *Cryptosporidium parvum* and *Giardia lamblia* (as low as 10 cysts or oocysts; Table 3-1).

**Selected Key Pathogens Associated with Livestock**

| Pathogen | Cattle | Poultry | Swine |
|---|---|---|---|
| *E. coli O157:H7* | X | | X |
| *Salmonella spp.* | X | X | X |
| *Campylobacter spp.* | X | X | X |
| *Yersinia entercolitica* | | | X |
| *Listeria spp.* | | | X |
| *Cryptosporidium parvum* | X | | |
| *Giardia lamblia* | X | | X |
| Rotavirus | X | | X |
| Norovirus | X | | X |
| Hepatitis E virus | | | X |

Minimizing the potential for human illness from pathogens in manure requires understanding the survival characteristics of the various pathogens. Survival times in manure and in the environment can range from days to years depending on the pathogen, the medium, and environmental conditions. Among the common zoonotic pathogens, however, *Cryptosporidium* is noteworthy because of its persistence, resistance to disinfection, and the lack of treatment for the illness it causes. It has been the causative agent of several large outbreaks for which manure has been identified as a possible source. Less is known about virus survival, and continued research is needed on virus occurrence, survival, and transport in environmental media.

Because of the different survival capabilities of the various pathogens, different manure management methods may be needed depending upon the pathogens anticipated; this is an area where further research is warranted. Composting of manure, especially when properly aerated, is an effective management practice that can generate the heat needed to inactivate a number of pathogens, including *Salmonella*, *Campylobacter*, *E. coli*, and protozoa. Ultraviolet light promotes die-off, and spreading manure on the surface during land application can promote greater die off through exposure to UV light and desiccation, although the manure is more susceptible to mobilization via runoff. Additional discussion of management methods is provided in Chapter 8.

Transport of pathogens may occur via runoff, air deposition, or infiltration into soils. The likelihood of significant transport of pathogens in runoff is increased where soils have low permeability or moderate to high antecedent moisture conditions, temperatures are below freezing, there is tile drainage, the slope of the land is steep, and rainfall is intense. Timing of manure land application is an important factor in minimizing pathogen transport via runoff. For example, avoiding application on frozen or snow-covered ground, during early spring runoff, when the land is saturated, or when the forecast calls for sufficient precipitation to produce runoff will help minimize pathogen loadings to surface water (Olson 2001). Transport of microorganisms in runoff is more likely if excess manure is applied or if manure is misapplied (USEPA 2002a). Once runoff reaches surface water bodies, microbes may become associated with bottom sediments if they are adsorbed to particles large enough to settle. Pathogens can, however, be reintroduced to the water column by resuspension after heavy rain events or human activities such as dredging.

During infiltration through soil, the upper layers of soil generally provide some removal of microbes through adsorption. The possibility of removal during transport through soil depends upon hydraulic conditions, soil texture and structure, soil composition, soil water composition, and microbial size and properties. Ground water is most vulnerable to contamination when manure is applied before a heavy rainstorm in an area with coarse, sandy soil and a shallow water table. Clayey soils may also promote transport to ground water if they have macropores and root channels.

# 4. Antimicrobials in Manure

Livestock and poultry are often given antimicrobials (i.e., antibiotics and vaccines) to treat and prevent diseases, as well as to promote animal growth and feed efficiency. Many of the antimicrobials administered to livestock and poultry are also used in human clinical medicine. Research indicates that sub-therapeutic use of antimicrobials can select for antibiotic resistance in bacteria. The purpose of this chapter is to provide estimates of the quantity and types of antimicrobials administered to livestock and poultry, and on aquaculture operations. Section 6.3 is a follow-up to this chapter, providing information on the extent of, and potential risks associated with, antimicrobial resistance related to livestock antimicrobial use.

## 4.1. Introduction

Antimicrobials have been administered to livestock and poultry for over 60 years (Libby and Schaible 1955). At therapeutic doses, antimicrobials help treat and prevent diseases and outbreaks. Administering antimicrobials at sub-therapeutic levels can enhance nutrient adsorption and limits the growth of microorganisms that may compete for nutrients, allowing the animal to grow to market weight more quickly, with less feed (MacDonald and McBride 2009).

Approximately 60% to 80% of livestock and poultry routinely receive antimicrobials through feed or water, injections, or external application (NRC 1999, Carmosini and Lee 2008). The majority of the antimicrobial use is estimated to be used for animal growth rather than for medicinal reasons, and many of these medications are also used in human clinical medicines (Mellon et al. 2001). Estimates suggest that as many as 55% of antimicrobial compounds administered to livestock and poultry are also used to treat human infections (Table 4-1) (Benbrook 2001, Kumar et al. 2005, Lee et al. 2007). The sub-therapeutic use of antimicrobials in livestock and poultry can facilitate the development and proliferation of antimicrobial resistance (Sapkota et al. 2007). Additionally, according to Boxall (2008) and Zounková et al. (2011), antimicrobials and their biologically active degradates may be discharged to the environment from livestock and poultry manure or, in the case of aquaculture, discharged directly to surface waters, potentially impacting aquatic life. The overlap between livestock and human antimicrobial use has been noted by the WHO and others as an area of concern for human health, because the effectiveness of these medications in treating human infections may be compromised (WHO 2000, Levy and Marshall 2004, Sapkota et al. 2007).

> ✓ Over 29 million pounds of antimicrobials were sold for livestock use in 2010 in the US – an estimated 3 to 4 times more than the amount used by humans.
>
> ✓ 60% to 80% of livestock routinely receive antimicrobials, the majority of which are estimated to be used for animal growth, rather than for medicinal purposes.
>
> ✓ The WHO has noted that sub-therapeutic antimicrobial use by livestock and poultry is an area of concern because of the selection for antimicrobial resistance.
>
> ✓ Antimicrobials generally do not biodegrade easily and may be more mobile in aquatic environments.

**Table 4-1. Select antimicrobials that are approved for use by the U.S. Food and Drug Administration for use in humans, livestock, and poultry.**

| Class/Group | Antimicrobial | Humans | Beef Cattle | Dairy Cows | Swine | Poultry | Aquaculture |
|---|---|---|---|---|---|---|---|
| Aminocyclitol | Spectinomycin | X | X | X | X | X | |
| Aminoglycoside | Apramycin | X | | | X | | |
| | Gentamicin | X | X | | X | X | |
| | Neomycin | X | X | X | X | X | X |
| | Streptomycin | X | X | X | X | X | |
| β-lactam | Amoxicillin | X | X | X | X | | |
| | Ampicillin | X | X | | X | | |
| | Cloxacillin | X | X | X | | | |
| | Penicillin | X | X | X | X | X | |
| Lincosamide | Lincomycin | X | | | X | X | |
| Macrolide | Erythromycin | X | X | X | X | X | |
| Polypeptide | Bacitracin | X | X | | X | X | |
| Polyene | Nystatin | X | | | | X | |
| Sulfonamide | Sulfadimethoxine | X | X | X | | X | X |
| Tetracycline | Oxytetracycline | X | X | X | X | X | X |
| | Tetracycline | X | X | X | X | X | |

*\*This table is not meant to be all-inclusive, and not all antimicrobials included in this table are listed in the individual livestock tables that follow. For a complete listing of antimicrobials approved for human and livestock use, visit the USFDA's website.*

## 4.2. Estimates of Antimicrobial Use

Quantifying livestock antimicrobial use is challenging and estimates vary widely because there are no publicly-available, reliable antimicrobial use data for food-producing animals (USGAO 2011a). Pharmaceutical companies are also not required to disclose veterinary drug sales information (Shore et al. 2009), and the types used at operations may be deemed proprietary information (Sapkota et al. 2007). Furthermore, use estimates based on dose rates can be complicated. While recommended antimicrobial doses for individual livestock and poultry range from 0.05 to 3.5 ounces per 1,000 pounds of feed (depending on the animal type and life stage), it is not uncommon for feed to contain more than the recommended dose (McEwen and Fedorka-Cray 2002, Kumar et al. 2005). For example, Dewey et al. (1997) reported that 25% of over 3,000 swine facilities studied in the U.S. supplied antimicrobials at concentrations greater than the recommended dose.

Estimating livestock and poultry antimicrobial use is also challenging because of the varying degrees of usage on different farms. For therapeutic applications, animals may be treated individually or as groups. Group application can be related to increased disease susceptibility in larger operations where livestock and poultry live in close confinement, facilitating infection and disease transfer (McEwen and Fedorka-Cray 2002, Kumar 2005, Becker 2010). In large livestock and poultry operations, antimicrobials may be administered to animals continuously or for extended periods of time at sub-therapeutic doses (e.g., in feed and water), because this approach is more efficient and sometimes the only feasible method of production (McEwen and Fedorka-Cray 2002). According to the USDA, 20% of swine feeder/finisher farms with less than 100 swine administered antimicrobials sub-therapeutically, whereas 60% of operations with 2,500 or more swine administered antimicrobials (MacDonald and McBride 2009). Antimicrobial use in aquaculture operations involves administration to the entire group by adding the antimicrobials directly to the water or via medicated feed pellets, which are added to the water (Zounková et al. 2011).

Recognizing the importance of quantifying livestock and poultry antimicrobial use, the U.S. Government Accountability Office (USGAO) has been advocating for better tracking and reporting mechanisms of antimicrobial use in livestock and poultry since 1999 (USGAO 2011a). In accordance with a 2008 amendment to the Animal Drug User Fee Act, the USFDA released estimates of the annual amount of antimicrobial drugs sold and distributed for use in livestock and poultry in 2009 and 2010 (USFDA 2010 and 2011a). The USFDA estimates that approximately 29.2 million pounds of antimicrobials were sold for livestock and poultry use in the U.S. in 2010 (USFDA 2011a), or a 62% increase over 1985 use estimates (U.S. Congress, OST 1995). Tetracyclines and ionophores were the largest class of antimicrobials reported, accounting for over 70% of all livestock and poultry antimicrobials sold during that year (USFDA 2011a). Overall, estimations of annual antimicrobial use in food animals in the U.S. range from 11 to 29.2 million pounds as reviewed in Table 4-2.

Given that many human health antimicrobials are also administered to livestock and poultry, and subtherapeutic use can select for resistance (Sapkota et al. 2007), it is important to understand the ratio between livestock and human antimicrobial use. The USFDA's (2010) reported sales of livestock and poultry antimicrobial use (approximately 28.8 million pounds in 2009) is estimated to be four times greater than what is used for human health protection (approximately 7.3 million pounds in 2009) (Loglisci 2010). A slightly higher ratio between livestock and human antimicrobial use was reported by Mellon et al. (2001), which estimated that livestock and poultry antimicrobial use in 1997 represented 87% of all antimicrobials used in the U.S.

The following subsections review antimicrobial use for cattle (beef and dairy), swine, poultry, and aquaculture to provide information on common diseases and infections that affect each animal type, and also provide estimates of the extent of antimicrobial use for therapeutic and sub-therapeutic purposes. Table A-10 in Appendix 2 summarizes animal life stages and definitions.

**Table 4-2. Estimates of antimicrobial use or sales for livestock in the U.S.**

| Total Mass Used/Sold | Specific Use | Source |
|---|---|---|
| 11 million pounds <u>sold</u> (in 1985) | Not Reported | Swartz 1989 |
| 18 million pounds used (in 1985) | 12.2% for treating disease<br>63.2% for disease prevention<br>24.6% for growth promotion | U.S. Congress, Office of Technology Assessment 1995 |
| 29.6 million pounds used (in 1997) | 7% for treating disease<br>93% for growth promotion and disease prevention | Mellon et al. 2001 |
| 17.8 million pounds used (in 1998) | 83% for prevention and treating disease<br>17% for growth promotion | Animal Health Institute 2000 |
| 28.8 million pounds <u>sold</u> (in 2009) | Not Reported | U.S. Food and Drug Administration 2010 |
| 29.2 million pounds <u>sold</u> (in 2010) | Not Reported | U.S. Food and Drug Administration 2011a |

*Adapted from Rogers and Haines (2005).*

### 4.2.1. Cattle (Beef and Dairy)

Beef cattle can be administered antimicrobials to treat or prevent common ailments such as respiratory disease (shipping fever and pneumonia), liver abscesses, bacterial enteritis (diarrhea), and coccidiosis (Table 4-3). Farming operations also administer prophylactic antimicrobials to beef cattle to promote feed efficiency and animal growth. An estimated 83% of beef cattle operations administered antimicrobials through animal feed or water for either animal growth or therapeutic purposes in 1999 (USDA 2000). During that same year, nearly all small (99%) and all large (100%) cattle feedlots used at least one parasiticide (USDA 2000). Parasiticides, such as ivermectin and doramectin, for example, are not antimicrobials but are used to kill parasites. A more recent USDA survey found that nearly 70% of beef cattle and calf operations vaccinated their animals and almost 70% of operations administered oral or injectable antimicrobials for disease treatment during 2007-2008 (USDA 2010b). Beef cattle operations with 200 or more cattle are more than twice as likely to vaccinate for bovine viral diarrhea virus (BVDV) than smaller operations with less than 50 cattle (USDA 2010b). Table 4-3 presents commonly used antimicrobials in beef cattle and their intended use.

**Table 4-3. Commonly used antimicrobials administered to beef cattle.**

| Class/Group | Antimicrobial | Life stage | Intended Use |
|---|---|---|---|
| Aminoglycoside | Gentamicin*, Neomycin*, Streptomycin* | Cattle | • Treat bacterial enteritis and pink eye |
| β-lactam | Amoxicillin*, Ampicillin*, Penicillin* | Cattle and calves | • Treat respiratory disease, bacterial enteritis, and foot rot<br>• Promote animal growth |
| Bambermycin | -- | Cattle (slaughter, feedlot) | • Promote feed efficiency and animal growth |
| Fluoroquinolone | Enrofloxacin | Cattle | • Treat respiratory disease |
| Ionophore | Lasalocid, Monensin | Unspecified | • Control coccidiosis<br>• Control liver abscesses<br>• Promote feed efficiency and animal growth |
| Macrolide | Erythromycin*, Tilmicosin, Tylosin | Calves | • Control calf diphtheria |
| | | Cattle | • Control metritis and liver abscesses<br>• Treat foot rot and respiratory disease<br>• Promote feed efficiency and animal growth |
| Polypeptide | Bacitracin* | Feedlot | • Control liver abscesses |
| | | Growing | • Promote feed efficiency and animal growth |
| Sulfonamide | Sulfamethazine | Calves | • Treat calf diphtheria |
| | | Cattle | • Treat respiratory disease, bacterial sores, foot rot, acute metritis, coccidiosis<br>• Promote animal growth in the presence of respiratory disease |
| Tetracycline | Chlortetracycline, Oxytetracycline* | Calves | • Treat bacterial pneumonia, bacterial enteritis, and diphtheria<br>• Promote feed efficiency and animal growth |
| | | Cattle | • Control liver abscesses and anaplasmosis<br>• Treat bacterial enteritis, foot rot, wooden tongue, and acute metritis<br>• Prevent bacterial pneumonia<br>• Promote feed efficiency and animal growth |

*(\*) indicates that the antimicrobial is approved for use in humans.*
*This table is meant to provide general antimicrobial use information. Antimicrobials listed within each class may be used for different purposes during particular animal life stages. Consult the USFDA's website for more specific information about livestock antimicrobial use. References: USGAO 1999, Herrman and Stokka 2001, McGuffey et al. 2001, Apley 2004, and USFDA 2011b.*

Similarly to beef cattle, dairy cows may be treated for respiratory disease and bacterial enteritis, but dairy cows may also be treated for other common ailments such as lameness and mastitis, which is a teat infection (Table 4-4; USDA 2008a). Most antimicrobials are prohibited for use on lactating cows when producing milk for human consumption (Watanabe et al. 2010). In 2007, 90% of dairy operations administered intramammary antimicrobials (e.g., lincosamide) during non-lactating periods, and 80% of those operations treated all cows at the facility (USDA 2008a). Approximately 85% of dairy operations used antimicrobials to treat mastitis, administering the antimicrobials to 16% of the cows on those operations (USDA 2008a). Preweaned heifers tend to be treated with antimicrobials more often than weaned heifers due to their increased susceptibility to diseases (USDA 2008a). Approximately 11% of preweaned heifers received antimicrobials to treat for respiratory disease, compared to 6% of weaned heifers (USDA 2008a). For growth promotion and disease prevention, 58% of dairy operations fed preweaned heifers dairy milk replacer, which was typically a combination of neomycin and oxytetracycline (USDA 2008a). In weaned heifers, approximately 45% of dairy operations used ionophores in feed for growth promotion and disease prevention (USDA 2008a).

**Table 4-4. Commonly used antimicrobials administered to dairy cows.**

| Class/Group | Antimicrobial | Life stage | Intended Use |
|---|---|---|---|
| Aminoglycoside | Neomycin*, Streptomycin* | Preweaned | • Treat bacterial enteritis and other digestive problems<br>• Promote animal growth |
| | | Unspecified | • Treat mastitis<br>• Prevent Staphylococcus aureus |
| β-lactam | Amoxicillin*, Cephalosporin, Penicillin* | Preweaned | • Treat bacterial enteritis and other digestive problems |
| | | Non-lactating | • Treat mastitis and lameness |
| | | Unspecified | • Treat respiratory disease and foot rot |
| Fluoroquinolone | Enrofloxacin | Non-lactating | • Treat respiratory disease |
| Ionophore | Lasalocid, Monensin | Weaned | • Treat for respiratory disease and bacterial enteritis<br>• Improved feed efficiency and growth promotion<br>• Increased milk production efficiency |
| Lincosamide | Pirlimycin Hydrochloride | Non-lactating | • Treat mastitis |
| Macrolide | Tilmicosin, Tylosin | Non-lactating | • Treat respiratory disease, foot rot, and metritis. |
| Sulfonamides | Sulfadimethoxine*, Sulfamethazine | Dairy calves and heifers | • Treat bacterial enteritis and other digestive problems<br>• Treat calf diphtheria, shipping fever complex, and foot rot |
| | | Non-lactating | • Treat acute mastitis and metritis |
| Tetracycline | Chlortetracycline, Oxytetracycline* | Preweaned | • Treat bacterial enteritis and other digestive problems<br>• Promote animal growth |
| | | Non-lactating | • Treat mastitis and lameness<br>• Treat bacterial enteritis and pneumonia |

*(\*) indicates that the antimicrobial is approved for use in humans.*
*This table is meant to provide general antimicrobial use information. Antimicrobials listed within each class may be used for different purposes during particular animal life stages. Consult the USFDA's website for more specific information about livestock antimicrobial use. References: USDA 2008a and USFDA 2011b.*

### 4.2.2.  Swine

Swine can be treated with antimicrobials to promote animal growth and to treat or prevent common infections such as respiratory diseases, swine dysentery, and bacterial enteritis (Table 4-5). According to the USDA, most hogs are raised in confinement, and large operations with 10,000 hogs or more typically administer antimicrobials through feed to promote animal growth, particularly in starter and grower hogs

(USDA 2002b, USDA 2008b). As with other types of livestock, antimicrobial administration varies by life stage (see Table 4-5). An estimated 89% of operations administer antimicrobials to grower/finisher pigs (hogs grown to market weight for slaughter) (USDA 2002b) and 85% of operations use antimicrobials in feed for nursery pigs (USDA 2008b). In the USDA (2008b) study, over half (54%) of the operations administered antimicrobials in the nursery pig feed continuously, while 33% of operations did so for grower/finisher pigs.

**Table 4-5. Commonly used antimicrobials administered to swine.**

| Class/Group | Antimicrobial | Life stage | Intended Use |
|---|---|---|---|
| Aminoglycoside | Gentamicin* | Preweaned | • Treat colibacillosis |
| β-lactam | Amoxicillin*, Ampicillin*, Penicillin* | Unspecified | • Promote feed efficiency and animal growth<br>• Treat bacterial enteritis, porcine colibacillosis, and salmonellosis |
| Bambermycin | -- | Growing/Finishing | • Promote feed efficiency and animal growth |
| Macrolide | Erythromycin*, Lincomycin, Tylosin | Starting/Growing/ Finishing | • Promote feed efficiency and animal growth<br>• Treat bacterial enteritis and infectious arthritis<br>• Control swine dysentery and the severity of swine mycoplasmal pneumonia |
| Pleuromutilin | Tiamulin | Unspecified | • Treat swine dysentery and pneumonia |
| Polypeptide | Bacitracin* | Growing/Finishing | • Promote feed efficiency and animal growth<br>• Control swine dysentery |
| Polypeptide | Bacitracin* | Pregnant | • Control clostridial enteritis |
| Tetracycline | Chlortetracycline, Oxytetracycline* | Growing | • Promote feed efficiency and animal growth<br>• Prevent/treat cervical lymphadenitis (jowl abscesses) |
| Tetracycline | Chlortetracycline, Oxytetracycline* | Breeding | • Prevent/treat leptospirosis |
| Tetracycline | Chlortetracycline, Oxytetracycline* | Unspecified | • Treat bacterial enteritis and pneumonia<br>• Reduce incidences of cervical abscesses |
| Streptogramin | Virginiamycin | Swine excluding breeders | • Promote feed efficiency and animal growth<br>• Treat swine dysentery |
| Sulfonamide | Sulfamethazine | Unspecified | • Promote feed efficiency and animal growth<br>• Control Bordetella bronchiseptica rhinitis<br>• Prevent swine dysentery and pneumonia<br>• Treat porcine colibacillosis and bacterial pneumonia |

*(*) indicates that the antimicrobial is approved for use in humans.*
*This table is meant to provide general antimicrobial use information. Antimicrobials listed within each class may be used for different purposes during particular animal life stages. Consult the USFDA's website for more specific information about livestock antimicrobial use. References: Herrman and Sundberg 2001, Mellon et al. 2001, Kumar et al. 2005, and USFDA 2011b.*

### 4.2.3.   Poultry

Poultry may be treated with antimicrobials to promote growth and to cure or prevent respiratory disease and infections, including *E. coli* and protozoan parasites such as coccidiosis (Table 4-6). The extensive use of antimicrobials in poultry, much of which is used for non-therapeutic purposes, has sparked consumer interest related to public health and antimicrobial resistance. For example, 3-Nitro (Roxarsone), the most commonly used arsenic-based drug for animals, promotes animal growth, improves pigmentation, and prevents coccidiosis in poultry (USFDA 2011c). In 2011, an USFDA study reported higher levels of inorganic arsenic (a known carcinogen) in broiler chickens treated with Roxarsone than non-treated broiler chickens, prompting the company producing the drug to suspend sales of Roxarsone for use in poultry (USFDA 2011c). Other arsenic-based drugs are still approved for use in poultry and swine, including nitarsone, arsanilic acid, and carbarsone (USFDA 2011c). In another instance, the use of fluoroquinolones in poultry was effectively banned by the USFDA in 2005 after research indicated an increase in human infections with fluoroquinolone-resistant *Campylobacter* related to poultry consumption (see Chapter 2 and Section 6.3 for further information) (Nelson et al. 2007).

**Table 4-6. Commonly used antimicrobials administered to poultry.**

| Class/Group | Antimicrobial | Life stage or Poultry Category | Intended Use |
|---|---|---|---|
| Aminocyclitol | Spectinomycin* | Chickens (not laying eggs for human consumption) | • Promote feed efficiency and animal growth<br>• Treat chronic respiratory disease<br>• Prevent mortality associated with Arizona group infection |
| Aminoglycoside | Gentamicin*, Neomycin* | Chickens and turkeys | • Prevent bacterial contamination and omphalitis<br>• Prevent early mortality caused by *E. coli* and *Salmonella typhimurium* |
| β-lactam | Penicillin* | Chickens/turkeys (not laying eggs for human consumption) | • Promote feed efficiency and animal growth |
| Bambermycin | -- | Broilers/growing turkeys | • Promote feed efficiency and animal growth<br>• Prevent coccidiosis<br>• Improve pigmentation |
| Ionophore | Lasalocid, Monensin | Broilers/turkeys | • Control of coccidiosis |
| Macrolide | Erythromycin*, Tylosin | Broilers/replacement chickens | • Control chronic respiratory disease |
| | | Layers | • Increase egg production |
| | | Chickens and turkeys | • Promote feed efficiency and growth promotion |
| Polypeptide | Bacitracin* | Broilers/replacement chickens | • Promote feed efficiency and animal growth<br>• Prevent necrotic enteritis |
| | | Layers | • Increase egg production<br>• Promote feed efficiency |
| | | Growing turkeys | • Promote feed efficiency and animal growth |
| Streptogramin | Virginiamycin | Broilers/turkeys | • Promote feed efficiency and growth promotion |
| Tetracyclines | Chlortetracycline | Chickens | • Promote feed efficiency and animal growth<br>• Control synovitis, chronic respiratory disease, air sac infections, and *E. coli* infections |
| | | Growing turkeys | • Promote feed efficiency and animal growth |
| | | Turkeys | • Control synovitis, hexamitiasis, and bacterial organisms associated with bluecomb |

*(\*) indicates that the antimicrobial is approved for use in humans.*
*This table is meant to provide general antimicrobial use information. Antimicrobials listed within each class may be used for different purposes during particular animal life stages. Consult the USFDA's website for more specific information about livestock antimicrobial use. References: Tanner 2000, McGuffey et al. 2001, Mellon et al. 2001, Apley 2004, Kumar et al. 2005, and USFDA 2011b.*

Estimates of antimicrobial use in poultry are limited. The 2010 poultry survey conducted by USDA's National Animal Health Monitoring System (NAHMS) program includes limited data on vaccine administration in breeder facilities, and no information is available on the types of drugs used or the extent of antimicrobial use in the poultry industry (USDA 2011a). According to the USDA's survey, in 2010, an estimated 80% of breeder chicken farms in the U.S. vaccinated pullets against *Salmonella*, bronchitis, and coccidiosis, among other infectious diseases (USDA 2011a). While the types of antimicrobials, including vaccines, were not reported in the USDA's poultry survey, as of 2009, at least 50 active pharmaceutical ingredients had been approved by the USFDA for use in poultry (USFDA 2009). Mellon et al. (2001) estimates that nearly 40% (10.5 million lbs.) of all antimicrobials used for non-therapeutic purposes in livestock and poultry during 1997 were administered to poultry. The study also suggests that the majority of poultry receive antimicrobials during at least one life stage. For example, layer eggs may be dipped in gentamicin to minimize bacterial contamination, and day-old chicks may be injected with gentamicin or other antimicrobials to prevent omphalitis, a yolk sac infection (Tanner 2000). Table 4-6 provides further information about commonly used antimicrobials in the poultry industry.

### 4.2.4.  Aquaculture

Antimicrobials may be used in aquaculture to prevent and treat bacterial infections and diseases (McEwen and Fedorka-Cray 2002). Primary antimicrobials used in aquaculture include oxytetracycline, sulfamerazine, sulfadimethoxine-ormetoprim combination, and formalin (Table 4-7). Estimates of total antimicrobial use in U.S. aquaculture vary widely. MacMillan et al. (2003) estimates that 54,000 to 72,000 pounds per year of antimicrobials are used in aquaculture, while Benbrook (2002) estimates that use is closer to 200,000 to over 400,000 pounds per year. Both estimates are significantly less than livestock and poultry antimicrobial use estimates; however, in contrast to livestock and poultry use, antimicrobials used in aquaculture enter surface waters directly, since they are added to the water through simple addition or via feed pellets (Lee et al. 2007, Zounková et al. 2011). Research suggests that, an estimated 70% to 80% of drugs administered in aquaculture operations are released into the environment, related to over-feeding and poor adsorption in the gut (Boxall et al. 2003, Gullick et al. 2007). As noted by Daughton and Ternes (1999) and Zounková et al. (2011), antimicrobials are designed to kill bacteria and may do so at multiple trophic levels, potentially impacting other, non-target, aquatic organisms. An assessment of the aquatic toxicity of 226 antimicrobials using USEPA's Ecological Structure Activity Relationships (ECOSAR) Class Program, predicted that a large portion of antimicrobials are toxic to aquatic life – algae, crustaceans, and fish (Sanderson et al. 2004). This is an area that needs further research.

**Table 4-7. Commonly used antimicrobials and parasiticides in aquaculture.**

| Class/Group | Antimicrobial | Life Stage or Species | Intended Use |
|---|---|---|---|
| Parasiticide (formaldehyde solution) | Formalin | Salmon, salmonids, and salmon eggs; trout and trout eggs; catfish, largemouth bass, bluegill, other fin fish, and shrimp | • Control of external protazoa, fungi, and protazoan parasites |
| Sulfanomide | Sulfadimethoxine*-Ormetoprim Combination, Sulfamerazine | Trout, salmonids, catfish | • Control furunculosis and enteric septicemia |
| Tetracycline | Oxytetracycline* | Salmonids, catfish, lobster | • Control ulcer disease, furunculosis, bacterial hemorrhagic septicemia, and pseudomonas disease |

*(\*) indicates that the antimicrobial is approved for use in humans.*
*This table is meant to provide general antimicrobial use information. Antimicrobials listed within each class may be used for different purposes during particular animal life stages. Consult the USFDA's website for more specific information about livestock antimicrobial use. References: Benbrook 2002 and USFDA 2011b.*

According to the USDA's 2005 Census of Aquaculture, catfish production is the dominant sector in U.S. aquaculture (USDA 2006). Approximately 50% of catfish hatcheries treated egg masses to control fungal and bacterial infections in 2009, with larger facilities more likely to administer antimicrobials than smaller ones (USDA 2010c). Additionally, approximately 29% of catfish fingerling operations administered antimicrobials in 2009 to treat and prevent enteric septicemia, a common bacterial infection in farm-raised catfish (USDA 2010c, USDA 2011b). Table 4-7 provides further information on antimicrobials used in aquaculture.

## 4.3. Antimicrobial Excretion Estimates

Antimicrobials are often only partially metabolized in livestock and poultry and can be excreted virtually unchanged as the parent compound (Kumar et al. 2005, Boxall 2008, Khan 2008, Pérez and Barceló 2008). For example, up to 80% of tetracyclines may be excreted by swine and poultry as the parent compound (Kumar et al. 2005, Khan 2008). Additionally, up to 67% of the macrolide tylosin, which is approved for use in beef cattle, dairy cows, swine, and poultry (see Table 4-3 to Table 4-6), may be excreted by livestock and poultry when the antimicrobial is administered orally (Feinman and Matheson 1978).

Several challenges are presented when attempting to estimate the types of antimicrobials present in livestock manure (i.e., dairy cow vs. beef cattle manure). First, as evidenced in the preceding tables (Table 4-3 to Table 4-7), the types of antimicrobials used at each operation differ depending on animal life stage and which ailments are most common at the operation. Second, dosage differs by operation, and excretion estimates vary by compound (McEwen and Fedorka-Cray 2002, Kumar et al. 2005). Finally, while hundreds of antimicrobial agents are approved for animal use, our understanding of which compounds are excreted is partly a function of which antimicrobials are tested for their presence in manure, as well as analytical detection limits. For example, Sapkota et al. (2007) estimated which antimicrobials to test for in ground water and surface water near a swine operation based on the types of antimicrobials approved for use by the USFDA. The actual antimicrobials used at the operation were deemed proprietary information, presenting a challenge to researchers in the environmental health field. Despite these limitations, recent research indicates that the most common antimicrobial classes found in manure include tetracyclines, macrolides, sulfonamides, ionophores, and β-lactams, some of which are also used for human health (Kumar et al. 2005, Lee et al. 2007).

## 4.4. Antimicrobial Stability and Transport in the Environment

After excretion, antimicrobials and their degradates can enter the environment in a variety of ways, including through direct land application via excretion from grazing animals or application of manure or lagoon slurry on cropland (Boxall 2008, Klein et al. 2008). Spills and overflow from manure lagoons, wash-off from indoor animal housing facilities or hard surfaces, and wash-off from animals treated externally also present pathways for antimicrobial transport to the environment (Boxall 2008, Klein et al. 2008). Additionally, antimicrobials can enter the atmosphere during the spraying of manure on fields, dust from scraping solid manure, or when antimicrobials bind to air particles during animal excretion (Boxall 2008, Chee-Sanford et al. 2009).

Antimicrobials are chemically diverse, though they tend to be hydrophilic and do not easily biodegrade; therefore these compounds tend to be more mobile in aquatic environments (Chee-Sanford et al. 2009, Zounková et al. 2011). However, because antimicrobials are organic compounds with a range of chemical properties, their stability and mobility in the environment varies considerably, with half-lives ranging from a few days to over a year (Kumar et al. 2005). Generally, antimicrobials tend to have a high affinity for soils and clays (Chee-Sanford et al. 2009). Tetracyclines, fluoroquinolones, and lincosamides are not considered to be very mobile related to their high sorption potential, while sulfonamides appear to be the most mobile of antimicrobials (Chee-Sanford et al. 2009). Antimicrobials with a high sorption potential may be less mobile in the environment, potentially persisting in cropland soil or at the bottom of manure lagoons for longer periods of time (Boxall et al. 2003, Lee et al. 2007, Adams et al. 2008, Carmosini and Lee 2008). Additionally, environmental factors such as pH, temperature, oxygen availability, and microbial populations can influence antimicrobial behavior and degradation in the environment (Gu and Karthikeyan 2005, Kumar et al. 2005, Carmosini and Lee 2008). Antimicrobials tend to degrade during manure storage, and the process appears to be more rapid under higher temperatures and aerobic conditions (Kumar et al. 2005, Lee et al. 2007, Boxall et al. 2008). Therefore, prolonged manure storage and avoiding manure land application during colder winter months may allow for further degradation, potentially reducing antimicrobial transport to the environment and surface waters. Given the limited number of field studies, further research in this area is warranted to determine optimal conditions for antimicrobial degradation in manure.

The majority of research on antimicrobial stability in the environment has been conducted in controlled laboratory experiments (Kumar et al. 2005, Lee et al. 2007). Some researchers are concerned that findings from these studies may not be directly applicable to actual conditions in the field since environmental factors, such as temperature and pH, fluctuate both spatially and temporally, influencing the behavior of antimicrobials in the environment (Sarmah et al. 2006). Further research on antimicrobial excretion and degradation in differing medias, including manure, soil, and water, may help researchers better quantify the amount of antimicrobials that enter the environment each year.

## 4.5. Antimicrobial Occurrence in the Environment

The occurrence of antimicrobials in soils, sediment, surface water, and ground water has been documented, particularly in close proximity to livestock and poultry operations. Campagnolo et al. (2002) found antimicrobial compounds present in 67% of ground water and surface water samples collected near poultry operations and 31% of ground water and surface water samples collected near swine operations. In that study, Campagnolo et al. (2002) detected lincomycin, chlortetracycline, and sulfadimethoxine, among other antimicrobials near both the swine and poultry operations. In another study, tetracyclines were detected in soils, and sulfonamides were detected in shallow ground water near large dairy livestock production facilities, which, in general, use significantly fewer antimicrobials per unit animal weight than other large livestock and poultry production facility types since most antimicrobials are prohibited for use on lactating cows (Watanabe et al. 2010). Additionally, Batt et al. (2006) detected two types of sulfonamides, which are approved only for veterinary use, in private drinking water wells near a large beef cattle livestock production facility and irrigated agriculture fields in Idaho. Lincomycin was measured in a ground water well near a swine lagoon in North Carolina (Harden 2009). In a study of North Carolina drinking water systems, fluoroquinolones as well as sulfonamides, lincomycin, tetracyclines, and macrolides were the most frequently detected antimicrobials in source water (Weinberg et al. 2004). In addition to livestock wastes, suspected sources also included wastewater treatment plants.

The concentrations of antimicrobials measured in the environment vary considerably, ranging from non-detectable concentrations to levels in the mg/L range. Overall, concentrations in soil tend to be much higher than in water because most antimicrobials bind well to soil (Lee et al. 2007). However, because antimicrobials tend to be hydrophilic, they can be transported in aquatic systems (Chee-Sanford et al. 2009, Zounková et al. 2011). It is important to note that our understanding of the occurrence of antimicrobials in the environment is limited by the fact that research tends to focus on the most commonly used antimicrobials (e.g., tetracyclines, sulfonamides), rather than degradates and less commonly used compounds. Numerous antimicrobial agents have been approved for livestock use, though many have not yet been researched in terms of their prevalence in the environment.

## 4.6. Summary and Discussion

Antimicrobial use is widespread in livestock and poultry production – both to treat infections and diseases, and also to increase feed efficiency and animal growth. An estimated 60% to 80% of livestock and poultry routinely receive antimicrobials (NRC 1999, Carmosini and Lee 2008), and several USDA surveys and publications suggest that larger, confined livestock and poultry operations rely more heavily on antimicrobial use than smaller facilities (MacDonald and McBride 2009, USDA 2010b). There are currently no reporting requirements for antimicrobial use on livestock and poultry operations, though according to the USFDA, an estimated 29.2 million pounds of antimicrobials were sold for livestock use in 2010 (USFDA 2011a). Gaining a more thorough understanding of the quantity of antimicrobials used in livestock and poultry production as well as the behavior and stability of antimicrobials in the environment may provide guidance for manure management to promote antimicrobial degradation prior to land application, thereby potentially reducing antimicrobial transport to the environment and surface waters. The possible link between livestock and poultry antimicrobial use and the proliferation and evolution of antimicrobial resistance (WHO 2000, Swartz 2002, USGAO 2011a) is discussed in Section 6.3.

This page intentionally left blank.

# 5. Hormones in Manure

Hormones are endocrine disruptors that are naturally produced by, and in some cases artificially administered to, livestock and poultry. As with all mammals including humans, livestock and poultry excrete hormones in their waste, which has the potential to enter water resources through runoff and discharges from animal production facilities and fertilized cropland. The purpose of this chapter is to provide estimates of livestock and poultry hormone use and excretion rates as well as the occurrence and mobility of hormones in the environment. Section 6.4 provides information on endocrine disruption and potential impacts to aquatic life and human health.

## 5.1. Introduction

Hormones are naturally synthesized in the endocrine systems of all mammals and regulate metabolic activity and developmental processes. Beef cattle may also be administered additional natural and synthetic exogenous hormones to improve beef quality and promote animal growth. Dairy cows may be treated with additional hormones to control reproduction and increase milk production (USFDA 2002, Bartelt-Hunt et al. 2012). The USFDA has not approved the use of exogenous steroid hormones for growth promotion purposes in swine, poultry, veal calves, or dairy cows (USFDA 2011d). Natural hormones include estrogens, androgens, and progestogens (Table 5-1), and their synthetic versions include zeranol, trenbolone acetate, and melengestrol acetate (Table 5-2).

> ✓ Livestock excreted an estimated 722,852 pounds of endogenous hormones in 2000.
>
> ✓ Beef cattle feedlot operations may administer synthetic hormones as implants and feed additives to promote animal growth.

**Table 5-1. Natural hormones and select metabolites as well as the functional purpose of the hormone.**

| Hormone | Select Hormone Metabolites | Purpose |
|---|---|---|
| Estrogens | Estrone, 17β-estradiol, and estriol | • Natural reproductive hormone<br>• Stimulates and maintains female characteristics |
| Androgens | Testosterone, 5α-dihydrotestosterone, 5α-androstane-3β, 17β-diol, 4-androstenedione, dehyroepiandrosterone, and androsterone | • Natural reproductive hormone<br>• Stimulates and maintains male characteristics |
| Progestogens | Progesterone | • Natural reproductive hormone<br>• Produced during the estrous cycle<br>• A metabolic precursor to estrogens |

Hormones are naturally excreted by livestock and poultry in manure and bile (USEPA 2004a, Zhao et al. 2008). Therefore, hormones and their metabolites can enter aquatic ecosystems through runoff from pasture and rangeland used by grazing cattle and cropland fertilized with manure, as well as via leaks/overflow from manure lagoons (Kolodziej and Sedlak 2007, Bartelt-Hunt et al. 2012). Because hormones are endocrine disrupting compounds, Lee et al. (2007) and Zhao et al. (2008), among others, have noted concern regarding the potential adverse impacts of aquatic organism exposure to manure. Specifically, hormones can affect the reproductive biology, physiology, and fitness of fish and other aquatic organisms (Zhao et al. 2008). It is important to note that all mammals excrete hormones, thus other possible sources of steroid hormones to the environment include wastewater treatment plant discharges and leaky septic systems (Shore and Shemesh 2003).

**Table 5-2. Synthetic hormones that may be administered to and excreted by beef cattle and/or dairy cows.**

| Synthetic Hormone | Mimics the Behavior of Which Natural Hormone Metabolite? | Purpose |
|---|---|---|
| Zeranol | 17β-estradiol | • Administered as an implant (typically without other hormones)<br>• Used to improve feed efficiency and animal growth |
| Trenbolone acetate | Testosterone | • Administered as an implant either alone or with 17β-estradiol<br>• Used to improve feed efficiency and animal growth |
| Melengestrol acetate | Progesterone | • Administered as a feed additive<br>• Used for estrous synchronization and to induce lactation<br>• Used to improve feed efficiency and animal growth |

## 5.2. Estimates of Exogenous Hormone Use

The USFDA has approved the use of patented forms of natural hormones and synthetic steroid hormones for use in beef and dairy cattle, as included in the Code of Federal Regulations (CFR), Title 21, Parts 522, 556, and 558 (see also Table 5-1 and Table 5-2). Hormones may be administered through implants, or pellets containing doses of one or more hormones that are implanted into the ear of an animal (USFDA 2011d). Typical implants on beef cattle feedlots contain doses of approximately 140 mg of trenbolone acetate and 14 mg of 17β-estradiol benzoate (Bartelt-Hunt et al. 2012). Beef cattle on feedlots may also receive daily doses of approximately 0.45 mg of melengestrol acetate in feed (Bartelt-Hunt et al. 2012). Intravaginal controlled internal drug release (CIDR) inserts, which contain progesterone, may be used in dairy operations to control estrous (menstrual cycle), or to treat anestrous (non-menstruating) females and females with cystic ovaries (USDA 2009c).

The USFDA has also approved the use of the genetically engineered hormone, recombinant bovine growth hormone (rBGH), also referred to as recombinant bovine somatotropin, to increase milk production in dairy cows (USFDA 2011e). Estimates of rBGH use in dairy cows are unknown; however, a 2006 USDA article reported that 33 million doses are sold annually by the manufacturer (Gray 2006) (note that this estimate may include sales outside of the U.S.). Information on the extent of rBGH treatments at U.S. dairy operations would allow for an understanding of trends in usage.

Estimates of hormone use in beef and dairy cattle are limited because there are no reporting requirements; however, recent USDA NAHMS surveys have provided insight into common practices in beef and dairy operations. Approximately 39% of steers and heifers weighing less than 700 pounds and 82% of those weighing 700 pounds or more received at least one hormonal implant in 1999 (USDA 2000). Of those, livestock operations with 8,000 or more cattle were more likely to use implants than smaller ones. Additionally, approximately 33% of dairy operations used CIDR inserts in 2007 (USDA 2009c). The USDA's NAHMS 2007 Dairy Survey mentions that rBGH is the most common production enhancement injection used in dairy operations, though use estimates are not provided (USDA 2009d). Beyond these estimates, research to-date (though limited) has focused primarily on livestock and poultry excretion, since hormones are also produced naturally, and use estimates therefore would not necessarily accurately reflect amounts entering the environment.

## 5.3. Hormone Excretion Estimates

Approximately 2.2 billion cattle, swine, and poultry generated an estimated 1.1 billion tons of manure in 2007 (see Chapter 2), and livestock excrete hormones that are naturally-produced and synthetic (in the case of cattle). Quantifying the total amount of hormones excreted by livestock and poultry is challenging because daily excretion rates vary by animal type, season, diet, age, gender, breed, health status, reproductive state, and whether or not the animal is castrated (Schwarzenberger et al. 1996, Lange et al. 2002, Khan et al. 2008). One of the most extensive estimates of hormone excretion currently available suggests that cattle, swine, and poultry (excluding turkeys), excreted approximately 722,852 lbs. of estrogens, androgens, and progestogens (excluding synthetic hormones) during the year 2000 (Table 5-3) (Lange et al. 2002). Cattle account for the majority of estrogen and progestogen excreted by livestock (93% and 92%, respectively), related to differences in excretion rates and the higher quantity of manure generated by cattle compared to other animal types. Androgens are predominantly excreted by cattle and poultry, followed by swine. Lange et al. (2002) estimate that adding excretion of exogenous hormones to the above figures may increase the total excretion values by as much as 0.2% for estrogens and 20% for androgens. Using these estimates, livestock excreted an estimated 724,900 lbs. of hormones in 2000 (an approximate 0.3% increase over the estimates in Table 5-3).

**Table 5-3. Estimated livestock and poultry endogenous hormone excretion in the U.S. in 2000.**

| Animal Type | Estrogens | | Androgens | | Progestogens | | Total | |
|---|---|---|---|---|---|---|---|---|
| | Lbs. | % of Total | Lbs. | % of Total | Lbs. | % of Total | Lbs. | % of Total |
| Cattle | 99,208 | 92.7% | 4,189 | 43.7% | 557,770 | 92.0% | 661,166 | 91.5% |
| Swine | 1,830 | 1.7% | 772 | 8.0% | 48,502 | 8.0% | 51,103 | 7.1% |
| Poultry (broilers, layers) | 5,952 | 5.6% | 4,630 | 48.3% | -- | -- | 10,582 | 1.5% |
| *Total* | *106,990* | *100%* | *9,590* | *100%* | *606,271* | *100%* | *722,852* | *100%* |

*(--) indicates that no estimate is available from Lange et al. (2002). Adapted from Lange et al. (2002).*

The following subsections provide information on hormone excretion rates for different animal types and aquaculture. Overall, limited data are available on hormone excretion, particularly for swine and poultry, and few studies have investigated aquaculture hormone contributions. Also, the majority of research has focused on estrogen excretion and, to a lesser extent, androgen excretion. Limited information is available on livestock progesterone and synthetic hormone excretion. Importantly, identifying trends and comparing data between livestock types is difficult because hormone excretion rates vary depending on the animal type and life stage.

### 5.3.1. Cattle (Beef and Dairy)

Hormone excretion in cattle varies by life stage and reproductive state, among other factors. For example, androgen excretion ranges from 0.0003 lbs./yr (120 mg/yr) in calves to 0.001 lbs./yr (390 mg/yr) in bulls (Lange et al. 2002). The majority (58% to 90%) of estrogen excreted by cattle is via feces, most of which is excreted during the final three months of pregnancy (Ivie et al. 1986, Lange et al. 2002, Shore et al. 2009). While pregnant cows produce significantly more hormones than non-pregnant cows, mean estrogen excretion rates within the first 80 days of pregnancy (first trimester) are similar to those of non-pregnant cattle (Hoffman et al. 1997). Pregnant cattle are estimated to excrete 0.01 lbs./yr (4,400 mg/yr) of progestogens (Lange et al. 2002).

Regarding excretion of synthetic, exogenous hormones, an estimated 8% of applied trenbolone acetate may be recovered in heifer liquid manure, and 3% to 42% may be recovered in solid dung (feces and straw)

(Schiffer et al. 2001). An estimated 12% of applied melengestrol acetate is excreted by heifers via feces (Schiffer et al. 2001). Limited information is available on zeranol and rBGH hormone excretion.

### 5.3.2. Swine

In contrast to cattle, which excrete the majority of total estrogen in feces, swine excrete nearly 96% of total estrogen in urine (Palme et al. 1996). Estrogen concentrations in swine manure tend to increase after three to four weeks of pregnancy (Choi et al. 1987, Szenci et al. 1997). Progestogen excretion can be as high as 0.009 lbs./yr (3,900 mg/yr) for pregnant swine, and 0.004 lbs./yr (1,700 mg/yr) for pigs in estrous (Lange et al. 2002).

### 5.3.3. Poultry

Similar to swine, the majority (69%) of total estrogen released into the environment by poultry is excreted via urine rather than feces (Ainsworth et al. 1962). Layers generally excrete more estrogen than broiler hens: 0.000016 lbs./yr (7.1 mg/yr) compared to only 0.00000075 lbs./yr (0.34 mg/yr) from broiler hens (Lange et al. 2002). Broilers generally excrete fewer androgens than laying hens and cocks. Androgen excretion by broilers is estimated to be 0.0000015 lbs./yr (0.7 mg/yr), while laying hens excrete 0.0000075 lbs./yr (3.4 mg/yr) and cocks excrete 0.0000196 lbs./yr (8.9 mg/yr) (Lange et al. 2002).

### 5.3.4. Aquaculture

As with mammals, fish and other aquatic organisms also naturally excrete hormones, though hormone contributions from aquaculture operations have been far less studied than livestock. Kolodziej et al. (2004) estimates that hormone discharge from a standard aquaculture operation (i.e., 55 to 220 tons of fish) may be comparable to the amount of hormones produced by several hundred cattle, *or* a wastewater treatment plant serving several thousand people. Hormone excretion may be higher during spawning periods, though further research is needed. In a study of hormone concentrations in aquaculture operations, Kolodziej et al. (2004) found that concentrations of estrone, testosterone, and androstenedione (a precursor to sex steroid hormones) ranged from 0.1 to 0.8 ng/L in hatchery effluents. Note that the rate of effluent production was not reported in the Kolodziej et al. (2004) study; therefore an estimate of hormone production reported as mass per year, cannot be calculated for these hatcheries. Effluent from aquaculture operations may enter natural surface waters untreated, either through direct discharge or overflow (Kolodziej et al. 2004).

## 5.4. Hormone Stability and Transport in the Environment

Because mammals, including livestock, poultry, and humans, produce and excrete hormones, key sources of hormones to the environment include manure and bile from livestock and poultry operations as well as biosolids and discharges from wastewater treatment facilities. As previously discussed, manure and biosolids are often land applied, which can lead to concentrated releases of hormones and other compounds (e.g., nutrients, pathogens, and antimicrobials) to the environment (Bevacqua et al. 2011). Related to the typically higher total weight of manure compared to biosolids, as well as the more extensive treatment of biosolids, the contribution of hormones to the environment from manure compared to biosolids can be higher. A recent analysis estimated that poultry litter application to farmland in Maryland is nearly two times greater than biosolids application, contributing approximately two times more progesterone (35.27 lbs./yr versus 17.6 lbs./yr) and six times more estrone (24.3 lbs./yr versus 4.2 lbs./yr) to the environment (Bevacqua et al. 2011).

The occurrence and stability of hormones in the environment have only recently been investigated, partly related to improvements in laboratory methods allowing for the detection of hormones at low (ng/L) concentrations. However, available monitoring data indicate that hormones and their metabolites have been detected in the environment in close proximity to livestock and poultry operations and generally degrade at different rates depending on the media and environmental conditions. Both estrogens and testosterone may degrade to other compounds after excretion (Zhao et al. 2008). While estrogens may be degraded by biotic or abiotic processes under either aerobic or anaerobic conditions, a key route of degradation for testosterone is through microbial activity (Zhao et al. 2008). Limited information is available on progesterone degradation, though some studies indicate that they may be actively transformed by spores and vegetative cells of microorganisms in soil, as well as some fungi (Plourde et al. 1974, Pokorna and Kasal 1990).

Hormones are lipophilic (fat soluble) organic molecules that generally do not readily dissolve in water (Casey 2004, Arnon et al. 2008). Because of these characteristics, hormones tend to sorb to sediment, soil particles, and organic matter (Arnon et al. 2008). Sorption potential measures how tightly the compound binds with soil particles and can thus be an indication of how likely the compound will leach from the soil. In a study of soil sorption potentials of estrogens in a range of soil types on cultivated land, Caron et al. (2010) found a significantly positive correlation between sorption potential and soil organic carbon content. While further research is needed, this finding suggests that hormone leaching and contributions to runoff may be minimized in soils with higher carbon content.

Hormones in the environment typically degrade over time. The extent and rate of degradation can depend on a variety of factors such as the media's moisture content, temperature, and organic carbon content, as well as the availability of light (Zhao et al. 2008). Microbial breakdown also appears to be a key route for the degradation of hormones; therefore, it is possible that hormones may persist for longer periods of time during colder, winter temperatures when microbial activity tends to be slower than during warmer months (Zhao et al. 2008).

**Table 5-4. Half-lives of natural and synthetic hormones in the environment.**

| Hormone (Metabolite) | Half-Life (days) | Media | Source |
|---|---|---|---|
| Estrogen (17β-estradiol) | 69 | Poultry manure compost | Hakk et al. 2005 |
| | 24 | Anaerobic soil | Ying and Kookana 2005 |
| | 0.2-9 | River | Jürgens et al. 2002 |
| Androgen (Testosterone) | 43 | Clay amended compost | Hakk et al. 2005 |
| Zeranol | 56 | Manure | USFDA 1994 |
| | 49-91 | Soil | USFDA 1994 |
| Trenbolone acetate | 267 | Liquid manure | Schiffer et al. 2001 |
| Trenbolone acetate (17α-trenbolone) | 0.2-2 | Aerobic soil | Khan and Lee 2010 |
| Trenbolone acetate (17β-trenbolone) | 0.2-.6 | Aerobic soil | Khan and Lee 2010 |
| Melengestrol acetate | 0.16-1 | Water | USFDA 1996 |

*Adapted from Zhao et al. (2008), Table 13.11.*

Manure storage may facilitate the degradation of natural and synthetic hormones. For example, the degradation of estrogen in manure during storage has been observed in broiler litter (Shore et al. 1995), manure from pregnant and non-pregnant cows (Schenkler et al. 1998), and dairy manure (Raman et al. 2001). However, research suggests that synthetic hormones may persist at low concentrations even after months of storage and land application. Schiffer et al. (2001) measured the fate of trenbolone acetate and melengestrol acetate in solid and liquid lagoon manure from cattle that had received hormone implants. Trenbolone acetate and melengestrol acetate were detected in the solid manure after excretion and also after 4.5 months of storage. Likewise, trenbolone was detected in the liquid manure, decreasing in concentration after 5.5 months of storage. However, trenbolone was still detected in the soil up to two months after the liquid manure was applied to corn fields and had an estimated half-life of 267 days during storage. As shown in Table 5-4, half-lives of natural and synthetic hormones vary considerably, ranging from several hours to over 260 days depending on the type of hormone and media.

## 5.5. Hormone Occurrence in the Environment

While limited, recent studies have detected hormones in manure, runoff, and in surface waters near livestock and poultry operations (e.g., Durhan et al. 2006, Kolodziej and Sedlak 2007, Bartelt-Hunt et al. 2012). However, analyzing trends and making definitive statements about hormone occurrence is challenging because many studies focus on the occurrence of one type of hormone or metabolite in one type of medium rather than researching the occurrence of an array of natural and synthetic hormones in the same study. Further, most studies involve the use of bioassay methods, which quantify total concentrations of $17\beta$-estradiol and testosterone; in contrast, chemical identification liquid chromatography-tandem mass spectrometry allows for more precise quantification of specific hormone compounds including estriol, $17\alpha$-estradiol and progesterone (Bevacqua et al. 2011).

Estrogen content in poultry litter (manure and bedding materials) is variable, ranging from 14,000 to 500,000 ppb ($\mu$g/kg) (Shore et al. 1993, 1995). Likely related to the higher portion of total estrogen that is excreted by poultry via urine (69%) rather than feces (Ainsworth et al. 1962), estrogen levels detected in dry broiler litter are substantially lower, at 28 ppb (Shore et al. 1995). The concentration of estrogen in manure from pregnant cows is around 36 ppb, with the estrogen content in bull manure estimated to be nearly four times lower (Shore 2009). The level of testosterone in dairy cow manure is estimated to be 25 ppb; concentrations in broiler litter vary from 30 to 133 ppb; in breeder layer litter, concentrations range from approximately 20 to 250 ppb (Shore et al. 1995, Lorenzen et al. 2004). The variability may be attributed to differences in breed, manure treatment, and age (Zhao et al. 2008). Progesterone levels in manure have been far less studied than other hormone compounds. However, Bevacqua et al. (2011) reported an average progesterone concentration of 63.4 ppb in poultry litter from 12 broiler chicken farms in the Mid-Atlantic.

Relatively few studies have focused on concentrations of synthetic hormones in manure, though a recent controlled experiment on feedlot beef cattle conducted by Bartelt-Hunt et al. (2012) provides insight into concentrations of synthetic hormones in manure. In that study, feedlot cattle were treated with exogenous hormones via implants and feed additives during two study seasons in 2007 and 2008. Average concentrations of melengestrol acetate ranged from 1.7 to 6.5 ppb in fresh manure, with concentrations generally decreasing from day seven of the study to day 109 (Bartelt-Hunt et al. 2012). The average concentration of $17\alpha$-trenbolone (a metabolite of trenbolone acetate) in fresh manure after 46 days was 31 ppb; average concentrations of $\alpha$-zearalanol and $\alpha$-zearalenol (metabolites of the synthetic hormone zeranol) were 47 ppb and 46 ppb respectively after 46 days.

Both natural and synthetic hormones and their metabolites have also been measured in runoff from livestock and poultry operations. Runoff from a Nebraska beef cattle feedlot with hormone-treated cattle had concentrations of testosterone of up to 420 ng/L, $17\alpha$-estradiol up to 720 ng/L, and estrone up to 1050 ng/L (Bartelt-Hunt et al. 2012). In another study, concentrations of $17\alpha$-trenbolone were detected in 67% of runoff samples from a beef cattle feedlot in Ohio with concentrations ranging from <10 to approximately 120 ng/L (Durhan et al. 2006).

A USGS nationwide reconnaissance survey of streams known, or suspected to be, susceptible to human, animal, or industrial impacts, reported that nearly 6% of streams had measureable concentrations of $17\alpha$-estradiol, with a median concentration of 30 ng/L (Kolpin et al. 2002). According to Hanselman et al. (2003) and Kolodziej and Sedlak (2007), the source of $17\alpha$-estradiol is likely cattle operations, given that this steroid is predominantly excreted by cattle and not by other types of livestock or humans. Shore et al. (1995) reported concentrations of up to 5 ng/L of estrogen and 28 ng/L of testosterone in small streams draining fields which had recently been fertilized with poultry litter. Runoff from cattle grazing rangeland may also contribute hormones to surface waters. Kolodziej and Sedlak (2007) detected steroid hormones in 86% of samples from rangeland creeks where cattle had access to the creeks. Though few studies are available, hormones have also been detected in ground water impacted by dairy farms (Arnon et al. 2008) and swine CAFOs (Harden et al. 2009). Concentrations of estrone and 17β-estradiol have been detected in manure storage ponds, with higher concentrations at increasing depths (Raman et al. 2004), and testosterone and estrogen have been detected in sediments below a dairy wastewater lagoon at depths of up to 148 ft and 105 ft, respectively (Arnon et al. 2008). Few studies have investigated the presence and stability of progesterone in the environment, though Zheng et al. (2008) found that progesterones were present in dried manure piles on a dairy operation, but not in dairy lagoon samples.

## 5.6. Summary and Discussion

Hormones are naturally synthesized by all mammals, including livestock and poultry. Estimates suggest that over 720,000 lbs. of natural and synthetic hormones were excreted in manure and bile by cattle, swine and poultry (excluding turkeys) in 2000 (Lange et al. 2002) (Table 5-3). Research (while limited) indicates that hormones and their metabolites may be present in the environment proximal to livestock and poultry operations, including streams, creeks draining cattle grazing rangeland, and surface waters downstream from beef cattle feedlots (Kolpin et al. 2002, Durhan et al. 2006, Kolodziej and Sedlak 2007, Arnon et al. 2008, Harden et al. 2009, Bartlet-Hunt et al. 2012). While hormones are typically detected at low concentrations, such chemicals are biologically active at low levels (ng/L) and are classified as endocrine disruptors (see Section 6.4). Manure storage prior to land application may promote hormone degradation (see Chapter 8), possibly minimizing the amount that enters the environment (Shore et al. 1995, Raman et al. 2001, Schiffer et al. 2001). However, the nature of the degradation products is not completely understood yet. More research on the use, occurrence, fate, and transport of natural and synthetic hormones is necessary in order to fully understand their potential impact on human and ecological health.

This page intentionally left blank.

# 6. Potential Manure-Related Impacts

Manure from livestock and poultry is a source of a number of contaminants including nutrients, pathogens, hormones, and antimicrobials (see Table 1-1). As reviewed in the previous chapters, these contaminants have been detected in manure and environmental media such as soil, sediment, and water resources near livestock and poultry operations. Manure can be viewed as a source of nutrients to water, and it may be related to the development of harmful algal blooms (HABs) in some cases. HABs can produce cyanotoxins – also contaminants of emerging concern. The purpose of this chapter is to review the potential and documented human health and ecological impacts associated with these contaminants. This is not a comprehensive discussion of human health issues related to manure and livestock and poultry operations. Additional health issues for people living in the vicinity of large animal feeding operations or working in livestock and poultry operations and handling manure are associated with air quality (see Donham et al. 2007, Merchant et al. 2005, Mirabelli et al. 2006, PCIFAP 2008).

## 6.1. Harmful Algal Blooms and Cyanotoxin Production

Nitrogen and phosphorus (nutrients) are perhaps the most widely researched pollutants from livestock and poultry manure. Nutrients from manure may reach surface water and ground water through runoff from pasture and cropland, infiltration through soil, or volatilization during manure decomposition leading to atmospheric deposition of nitrogen (Jordan and Weller 1996, Bouwman et al. 1997, Aneja et al. 2001). Nutrients are necessary for all biological growth, but excess nutrients may lead to eutrophication in aquatic ecosystems. Characterized in part by excessive algal growth and potentially harmful algae blooms (HABs), eutrophication can alter the biology, chemistry, and aesthetic quality of the waterbody. HABs can also produce toxins, which may be harmful to wild animals and aquatic life as well as to humans and pets when exposed to them from drinking water supplies or recreational waters (see Grand Lake St. Marys case study) (Lopez et al. 2008).

While livestock and poultry manure contributes nutrients to the environment, there have been limited cases where manure has been documented as the primary cause of HABs and associated formation of cyanotoxins. Additionally, livestock and poultry manure must be placed in context relative to all the nutrients used in agricultural production. The National Research Council (NRC) estimated nitrogen and phosphorus

> **Manure-Related Harmful Algal Blooms in Grand Lake St. Marys, Ohio**
>
> Grand Lake St. Marys (GLSM) is a public drinking water supply in Ohio that has experienced recurring HABs since 2009 related to livestock manure runoff and nutrient loading (OEPA 2009). The watershed is 90% agricultural, with nearly 300,000 animal units of poultry, swine, and cattle. The HABs have caused fish kills, waterfowl and pet deaths, and have also been linked to over 20 cases of human illness. The state of Ohio has issued recreation, boating, and fish consumption advisories related to the blooms. The $150 million annual lake-based recreational and tourism industries have been compromised, park revenues have decreased by more than $250,000 per year, and several lakeside businesses have closed. To date, millions of state, federal, and local dollars had been leveraged toward lake restoration and watershed management projects. Technical assistance and funding programs have also been developed to minimize manure runoff to the lake. (References: OEPA 2007, OEPA 2009, OEPA 2011, Gibson 2011).

balances for croplands by USDA Region and for the U.S. The NRC reported that in the U.S., 45% of nitrogen and 79% of phosphorus inputs to cropland may be attributed to synthetic fertilizers, whereas 8% of nitrogen and 15% of phosphorus inputs are from livestock and poultry manure (NRC 1993). However, because manure production is more localized (refer to Chapter 2), associated nutrient contributions can be higher in particular watersheds. For example, a USGS study found that animal manure was the primary source of nitrogen in several Mid-Atlantic and southern watersheds, contributing 54% and 56% of total nitrogen loads to the Susquehanna River in Pennsylvania and the White River in Arkansas, respectively (Puckett 1994).

The majority of HABs in freshwater in the U.S. and throughout the world are caused by cyanobacteria, commonly referred to as blue-green algae. USEPA's 2007 National Lakes Assessment found that microcystin, a hepatotoxin produced by cyanobacteria that is harmful to animals and humans, was detected in approximately one third of the lakes studied (USEPA 2010b). It is important to note that the presence of cyanobacteria is not necessarily an indication of cyanotoxins because not all cyanobacteria, and not all blooms produce toxins. Table 6-1 reviews the various types of nuisance and harmful algae, the toxins they can produce, and the associated adverse human health and aquatic life impacts.

**Table 6-1. Types of harmful or nuisance inland algae, toxin production, and potential adverse impacts.**

| Algae Group | Genera/Taxa | Toxins | Potential Adverse Impacts |
|---|---|---|---|
| Cyanobacteria | Anabaena, Aphanocapsa, Hapalosiphon, Microcystis, Nostoc, Oscillatoria, Planktothrix, Nodularia spumigena, Aphanizomenon, Cylindrospermopsis, Lyngbya, Umezakia | Hepatotoxins, neurotoxins, cytotoxins, dermatoxins, endotoxins, respiratory and olfactory irritant toxins | • Human and animal health impacts (i.e., gastrointestinal disorders, liver inflammation/failure, tumor promotion, cardiac arrhythmia, skin irritation, respiratory paralysis, etc.)<br>• Water discoloration<br>• Unpleasant odors and aesthetics<br>• Hypoxia from high biomass blooms<br>• Taste and odor problems in drinking water and in farm-raised fish |
| Haptophytes | Prymnesium parvum, Chrysochromulina polylepis | Ichthyotoxins | • Fish mortalities |
| Chlorophytes, Microalgae | Volvox, Pandorina | -- | • Water discoloration<br>• Localized hypoxia |
| Macroalgae | Cladophora | -- | • Unpleasant odors and aesthetics<br>• Localized hypoxia<br>• Clogged water intakes |
| Euglenophytes | Euglena sanguinea | Ichthyotoxins | • Water discoloration<br>• Fish mortalities |
| Raphidophytes* | Chattonella | Ichthyotoxins | • Fish mortalities |
| Dinoflagellates | Peridinium polonicum | Ichthyotoxins | • Fish mortalities |
| Cryptophytes | Cryptomonas, Chilomonas, Rhodomonas, Chroomonas, Hemiselmis, Proteomonas, Teleaulax$^\Omega$ | -- | • Water discoloration<br>• Localized hypoxia |
| Diatom | Didymosphenia geminata | -- | • Produce large quantities of extracellular stalk material resulting in ecosystem and economic impacts |

*\* Raphidophytes are a marine algae, but can bloom in inland saline waters*
*$\Omega$ Information from Marin et al. (1998).*
*Adapted from Lopez et al. 2008.*

## 6.2. Fish Kills

Manure discharges to surface waters have been implicated in fish kills nationwide (Mulla et al. 1999). Such discharges can be caused by rain events, equipment failures (e.g., lagoon ruptures/leaks), or the application of manure to frozen ground or to tile drained fields, and subsequent discharges to surface waters. Fish mortalities from runoff containing manure may be caused by ammonia toxicity and/or oxygen depletion with large loadings of manure.

In Minnesota, a top swine producing state, an estimated 20 manure spills occur annually, one of which involved 100,000 gallons of liquid hog manure washing into Beaver Creek, killing nearly 700,000 fish (DeVore 2002). Similarly, in Lewis County, New York, millions of gallons of manure from a dairy CAFO spilled from a lagoon in 2005, contaminating approximately 20 miles of the Black River and killing approximately 375,000 fish (NYSDEC 2007). In 1995, spills from poultry and swine lagoons entered Cape Fear River basin in North Carolina, causing fish kills, algal blooms, and microbial contamination (Mallin and Cahoon 2003). Osterburg and Wallinga (2004) reported over 300 manure spills within ten years in Iowa alone, 24% of which were caused by manure storage overflow and equipment failures. Large livestock and poultry operations often store large volumes of untreated manure in lagoons, which can rupture or overflow, leading to a greater potential for fish kills (Armstrong et al. 2010). Between 1995 and 1998 alone, there were an estimated 1,000 manure spills at animal feedlots in ten states and 200 manure-related fish kills in the U.S. (Marks 2001). Proper management and maintenance of lagoons and minimization of winter land application of manure will help prevent manure discharges to surface waters.

## 6.3. Antimicrobial Resistance

Antimicrobials are typically administered to livestock therapeutically for disease treatment, control, and prevention, as well as sub-therapeutically for growth promotion (refer to Chapter 3) (Kumar et al. 2005). The USFDA estimates that 29.2 million lbs. of antimicrobials were sold for livestock and poultry use in 2010 (USFDA 2011a). The use of antimicrobials in livestock and poultry has been increasing over the past four decades (Pérez and Barceló 2008). This increase is partly related to the shift towards fewer, larger confined animal facilities, which may increase disease susceptibility among livestock because the livestock are routinely in close contact (Pérez and Barceló 2008). The overuse and/or misuse of antimicrobials (in general) can facilitate the development and proliferation of antimicrobial resistance (i.e., when bacteria have the ability to survive exposure to certain types of antimicrobials) (Levy and Marshall 2004). Research conducted by the WHO and others suggest that antimicrobial use in livestock and poultry, which is typically administered at low doses for extended periods of time for sub-therapeutic purposes, has contributed to the prevalence of antimicrobial-resistant pathogens found in food animal operations and nearby environments (WHO 2000, Swartz 2002, Hayes et al. 2004, Levy and Marshall 2004, Nelson et al. 2007, USGAO 2011a). However, antimicrobial resistance can develop in a number of ways, and while resistant infections in humans have been linked to livestock and poultry production (Swartz 2002), the relationship between livestock and poultry antimicrobial use and resistant infections in humans is not well understood. This section focuses on antimicrobial resistance and the potential human health implications. Note that research also indicates that antimicrobials are toxic to aquatic life; this topic has been reviewed elsewhere (e.g., Sanderson et al. 2004, Kümmerer 2009a and 2009b) and is not the focus of this chapter.

## 6.3.1.    Development and Spread of Antimicrobial Resistance

Each class of antimicrobials operates differently: some attack cell walls and membranes, some act on cellular components responsible for protein synthesis, and others interrupt biochemical pathways within the cell (Rogers and Haines 2005). Bacteria may develop resistance to antimicrobials when their deoxyribonucleic acid (DNA) changes through the mutation of existing genetic material. Bacteria may also develop resistance through conjugation (i.e., the transfer of genetic material between living bacteria), transformation (i.e., obtaining genetic material from the environment), or transduction (i.e., the transfer of genetic material between bacteria via a bacteriophage) (Rogers and Haines 2005). Because of the multiple methods by which resistance can spread, exposure of bacteria to increasingly large pools of antimicrobial resistant genes can further expand the pool of resistant strains of pathogens.

Antimicrobial-resistant bacteria are generally shed in animal manure, but they may also be present in the mucosa of livestock animals. Once a resistant strain is present in a bacterial community, it can spread among livestock, wild animals, pets, and humans (Figure 6-1). For example, resistance can spread between herds of animals, particularly when in close confinement, or via vectors such as insects and rodents (McEwen and Fedorka-Cray 2002). Antimicrobial-resistant pathogens can also survive on food products, such as vegetables and fruit grown on fields fertilized with manure containing resistant pathogens, or meat from slaughterhouses; such pathogens can also spread through soil or water that has been contaminated with manure containing resistant bacteria (USGAO 2011a). It is important to note that ingested bacteria will not always cause illness, in part because many strains of bacteria are naturally present in the human and/or animal digestive tract (e.g., certain strains of *E. coli*) (USGAO 2011a).

> ✓  The sub-therapeutic use of antimicrobials in livestock contributes to the development of antimicrobial resistant pathogens.
>
> ✓  The U.S. Department of Agriculture reported that 74% of *Salmonella* and 62% of *Campylobacter* isolates from swine manure were resistant to two or more antimicrobials.
>
> ✓  Resistant strains of pathogens tend to be less responsive to treatment and can cause more severe and prolonged illness in humans than susceptible strains.
>
> ✓  The U.S. Food and Drug Administration banned the use of fluoroquinolones in poultry in 2005 related to human health concerns; livestock antimicrobial use has previously been banned in European countries related to perceived human health concerns.

Most antimicrobial resistance related to human health is likely the result of overuse and misuse of certain medications in humans (Levy and Marshall 2004). However, evidence suggests that the use of antimicrobials in livestock and poultry operations selects for antimicrobial resistance in certain pathogens and bacteria such as *Salmonella* and *Enterococcus* (McEwen and Fedorka-Cray 2002). These bacteria may be transferred to humans through the food chain and via contaminated water (McEwen and Fedorka-Cray 2002).

**Figure 6-1. Potential pathways for the spread of antimicrobial-resistance from animals to humans.**

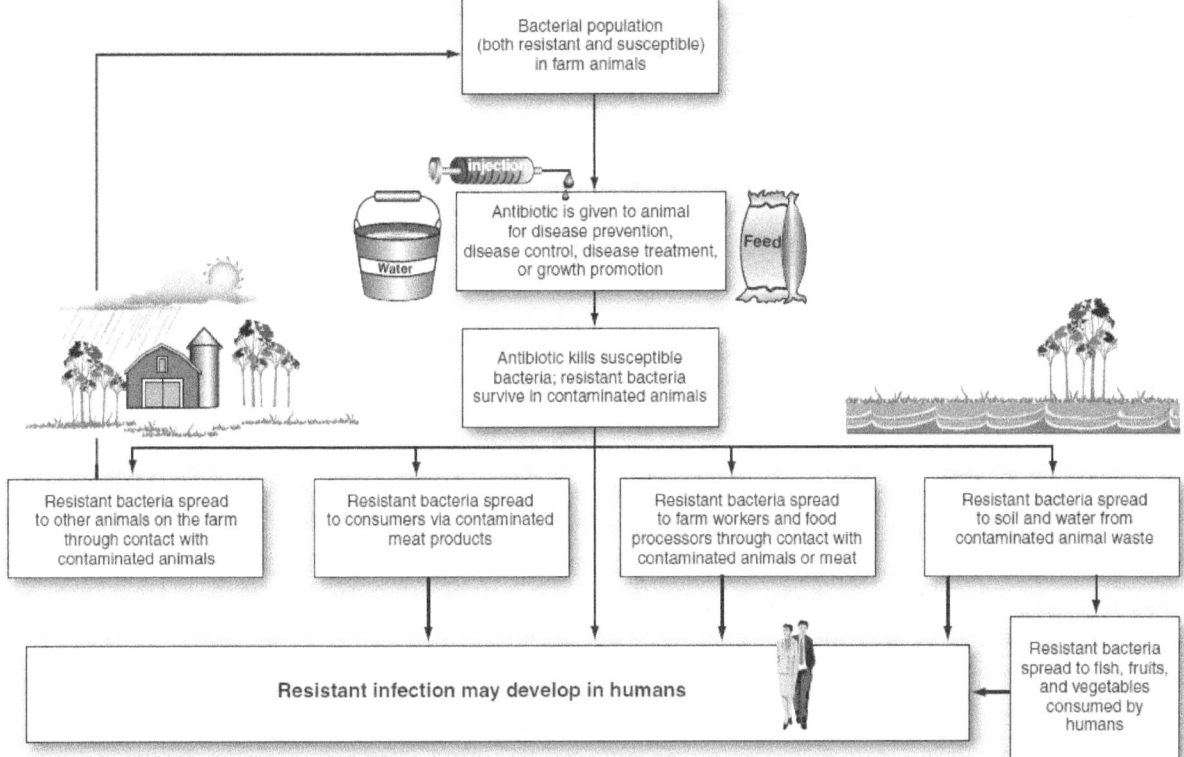

*\*As indicated in the figure, antimicrobial-resistant pathogens can spread to humans through several pathways. Certain pathogens with resistance can infect humans, increasing the severity and decreasing the treatability of the resulting illness/infection. Source: USGAO (2011a), Figure 1.*

### 6.3.2.   Antimicrobial Resistance in Manure and the Environment

Antimicrobial-resistant pathogen strains can be shed by livestock and poultry and are therefore generally found in manure and nearby environments such as surface water, ground water, and fertilized cropland. Antimicrobial-resistant *Enterococcus* spp. isolates were found to be prevalent in broiler and layer chicken operations in the Netherlands, with over 90% of isolates resistant to oxytetracyline or erythromycin (van den Bogaard et al. 2002). In that study, 80% of *Enterococcus* spp. isolates from broiler litter were also resistant to vancomycin, which is typically the first line drug used in humans to treat *Enterococcus* infections. Note that vancomycin has not been approved by the USFDA for use by livestock and poultry in the U.S. In a separate survey of poultry litter from more than 80 broiler operations, approximately 99% of *Enterococcus* spp. isolates were resistant to lincomycin, 68% were resistant to tetracycline, 54% were resistant to erythromycin, and 27% were resistant to penicillin (Table 6-2) (Hayes et al. 2004). Each of these medications is also used to treat human infections, and some may be used to treat infections from *Enterococcus*, specifically. Importantly, whether or not antimicrobial use in the poultry was a direct cause of the high prevalence of resistance is unclear because the types and quantities of antimicrobials used on the farms in the Hayes et al. (2004) study were not known/reported.

Research indicates that increased use of antimicrobials in livestock and poultry may be related to a greater prevalence of resistant pathogens in manure. Jackson et al. (2004) reported that 59% of *Enterococcus* spp. isolates were erythromycin-resistant in manure from a swine farm administering tylosin continuously through feed for animal growth, compared to 28% in a swine farm that administered tylosin for disease treatment for only five days (both tylosin and erythromycin are macrolides). The percent occurrence of erythromycin-resistant isolates was only 2% on a swine farm that did not use tylosin. Similarly, Sapkota et al. (2011) reported a significantly lower occurrence of antimicrobial-resistant strains of *Enterococcus* spp. on organic, antimicrobial-free poultry farms compared to conventional poultry operations. On the conventional operations, 42% of *Enterococcus faecalis* (*E. faecalis*) and 84% of *Enterococcus faecium* (*E. faecium*) isolates were multidrug-resistant (Table 6-2), compared with only 10% of *E. faecalis* and 17% of *E. faecium* isolates on the organic operations.

Results from USDA's NAHMS studies on the occurrence of antimicrobial-resistant pathogens in livestock and poultry manure, suggest a higher prevalence of antimicrobial-resistant pathogens in manure from swine, compared to other animal types (see USDA sources in Table 6-2). This finding was also reported by Sayah et al. (2005), which researched antimicrobial resistance patterns in livestock and poultry, companion animals, human septage, wildlife, surface water, and farm environments (e.g., manure storage facilities, lagoons, and livestock holding areas) in a watershed in Michigan. In that study, *E. coli* isolates from livestock manure were resistant to the greatest number of antimicrobials, and multidrug resistance was most common in isolates from swine manure (Table 6-2). Resistance was demonstrated most frequently to tetracycline, sulfisoxazole, streptomycin, and cephalothin (a type of cephalosporin that has since been voluntarily withdrawn from the U.S. market by the drug manufacturer). In terms of *Salmonella* and *Campylobacter*, the USDA's NAHMs studies also indicate that antimicrobial-resistant strains of these pathogens are less prevalent in beef cattle manure compared to dairy cow and swine manure (Table 6-2).

**Table 6-2. Occurrence of antimicrobial-resistant isolates in livestock and poultry manure from conventional livestock operations.**

| Pathogen | Animal Type | % of Resistant Isolates | Source |
|---|---|---|---|
| *Salmonella* spp. | Beef cattle | 0% resistant to any antimicrobials | USDA 2009e |
| | Dairy cows | 2% resistant to 1 antimicrobial<br>6% resistant to ≥ 2 antimicrobials | USDA 2009f |
| | Swine | 80% resistant to 1 antimicrobial<br>74% resistant to ≥ 2 antimicrobials | USDA 2009g |
| *Escherichia coli* | Swine | 32% resistant to 1 antimicrobial<br>60% resistant to ≥ 2 antimicrobials | USDA 2009h |
| | Swine | 31% resistant to 1 antimicrobial<br>15% resistant to > 3 antimicrobials | Sayah et al. (2005) |
| | Dairy cows | 28% resistant to 1 antimicrobial<br>6% resistant to > 3 antimicrobials | |
| | Beef cattle | 28% resistant to 1 antimicrobial<br>6% resistant to > 3 antimicrobials | |
| | Poultry (broilers) | 28% resistant to 1 antimicrobial<br>12% resistant to > 3 antimicrobials | |
| *Enterococcus* spp. | Poultry (broilers) | 53% resistant to 4 antimicrobials | Hayes et al. (2004) |
| | Poultry (broilers) | 42% (*E. faecalis*) resistant to ≥ 3 antimicrobials<br>84% (*E. faecium*) resistant to ≥ 3 antimicrobials | Sapkota et al. 2011 |
| *Campylobacter* sp. | Beef cattle | 8% resistant to ≥ 2 antimicrobials | USDA 2009i |
| | Dairy cows | 62% resistant to 1 antimicrobial<br>2% resistant to ≥ 2 antimicrobials | USDA 2009f |
| | Swine | 91% resistant to 1 antimicrobial<br>62% resistant to ≥ 2 antimicrobials | USDA 2008c |

Antimicrobial-resistant pathogens have also been detected in surface water and ground water near livestock and poultry operations. In the Sayah et al. (2005) study previously described, antimicrobial-resistant isolates of

*E. coli* were detected throughout the farm environment as well as in surface water near farming operations. Among the surface water samples, 81% of *E. coli* showed resistance to cephalothin (Sayah et al. 2005). Ash et al. (2002) reported that over 40% of bacteria in 16 rivers in the U.S. were resistant to at least one antimicrobial. Chee-Sanford et al. (2001) reported resistant bacteria in swine lagoons and underlying ground water, with the bacteria detected over 800 ft. down-gradient from the lagoons. In a study of the presence of resistant bacteria near a concentrated swine operation, median levels of enterococci and *E. coli* were up to 33 times higher in surface water and ground water down-gradient from the operation. A higher percentage of the enterococci were resistant to erythromycin and tetracycline in surface water samples, and a higher percentage of resistance to tetracycline and clindamycin were observed in down-gradient ground water samples. The surface water was used for recreational purposes, and the ground water had been used as a primary drinking water source but was taken offline due to pollution from the swine operation (Sapkota et al. 2007). The presence of antimicrobial-resistant bacteria in flowing systems such as streams, rivers, and ground water may facilitate the spread of resistant bacteria in the environment (McEwen and Fedorka-Cray 2002).

The presence of antimicrobial-resistant bacteria in drinking water source water and tap water has been documented. Bacteria resistant to amoxicillin, chloramphenicol, ciprofloxacin, gentamicin, sulfisoxazole, and tetracycline were found in surface water sources of drinking water in Michigan and Ohio (Xi et al. 2009). The percent of resistant bacteria ranged from 1.66% to 14.42% in source water, and from 1.17% to 47.98% in finished (treated) water. The study found that the levels of antibiotic-resistant bacteria were higher in tap water compared to finished water, suggesting that bacteria continued to grow in the drinking water distribution system (Xi et al. 2009).

The presence of antimicrobial-resistant bacteria in air, soil, and on cultivated land has also been documented. Gibbs et al. (2004) detected antimicrobial-resistant bacteria in air samples inside and downwind of a concentrated swine operation, but not upwind, suggesting that the swine operation was the source of the resistant bacteria. Multidrug-resistant bacteria have also been detected in topsoil from dairy farms, demonstrating resistance to chloramphenicol, penicillin, nalidixic acid, and tetracycline (Burgos et al. 2005). In soil from farmland amended with swine manure slurry, there was an increase in tetracycline-resistant bacteria following manure application, though the amount of resistant bacteria decreased during the eight months of the study (Sengeløv et al. 2003).

The period of time between antimicrobial introduction and the emergence of antimicrobial-resistant pathogens on a livestock operation varies. Because of the numerous ways in which bacteria can gain resistance (see subsection 6.3.1), once the pool of resistant genes reaches a certain magnitude, reversal of the problem can be challenging (Swartz 2002). While limited, available research suggests that certain antimicrobial-resistant pathogens may be more persistent in the environment than

---

**The USFDA Bans Prophylactic Use of Cephalosporin in Livestock**

Cephalosporins are antimicrobials used to treat pneumonia, pelvic inflammatory disease, and skin infections in humans. They are also widely used in livestock production; the USFDA reported that over 54,000 lbs. were sold for use in food-producing animals in 2010. Also, a USDA survey reported that in 2007, over half (53%) of dairy operations administered cephalosporins to treat mastitis (an increase from 37% of operations in 2002). There has been growing concern over the increased prevalence of cephalosporin-resistant pathogens (i.e., *Salmonella* and *E. coli*) related to widespread livestock use. To preserve the effectiveness of cephalosporins for human use, the USFDA has moved to ban their prophylactic use (among other uses) in cattle, swine, and poultry. The new rule became effective in April, 2012. (References: USDA 2008a, USFDA 2011a and 2012. Gilbert 2012).

---

others. However, research on the persistence of resistant pathogens appears to be focused primarily on *Campylobacter* and *Enterococcus* in the poultry industry, so there is a strong need for more research in this area.

Fluoroquinolone-resistant *Campylobacter* appears to be persistent in poultry operations. Price et al. (2005, 2007) researched the prevalence of resistant strains of *Campylobacter* in chicken meat products from two prominent poultry companies that had discontinued the use of fluoroquinolones in drinking water to treat entire flocks. In the study, even one year after discontinuing the use of the drug, fluoroquinolone-resistant *Campylobacter* was detected in 43% to 96% of the chicken products from the two producers. Chicken products from one of the producers were over 450 times more likely to carry fluoroquinolone-resistant *Campylobacter* than products from an antimicrobial-free poultry operation involved in the study (Price et al. 2005). There was no significant change in the proportion of resistant *Campylobacter* strains three years later (i.e., four years after the operations had discontinued the use of fluoroquinolones) (Price et al. 2007). The persistence of fluoroquinolone-resistant *Campylobacter* is of interest, because this pathogen is a primary cause of bacterial gastroenteritis in the U.S., causing approximately 1.4 million infections annually (Nelson et al. 2007). Fluoroquinolones are commonly prescribed to adults infected with *Campylobacter* (Nelson et al. 2007). Thus, resistance compromises the effectiveness of these antimicrobials in treating *Campylobacter* infections in humans. As described in subsection 6.3.3, the USFDA has since banned the use of fluoroquinolones in poultry due to fluoroquinolone resistance and human health concerns.

Research conducted in the U.S. and in Europe indicates that antimicrobial-resistant *Enterococcus* spp. may be less persistent than *Campylobacter*. For example, one study found that five newly organic and antimicrobial-free large-scale poultry operations in the U.S. experienced a substantial drop in the prevalence of antimicrobial-resistant *Enterococcus* spp. in feed, litter, and water samples, compared to five conventional operations (see subsection 6.3.2) (Sapkota et al. 2011). Similarly, tylosin-resistant *Enterococcus* spp. isolates detected in swine manure in Denmark were high (around 90% occurrence) prior to Denmark's ban of the use of tylosin for growth promotion (Aarestrup et al. 2000). However, the percent occurrence of tylosin-resistant *Enterococcus* spp. isolates decreased to 28% and 47% for *E. faecalis* and *E. faecium*, respectively, three years after the ban. It is important to note that a more substantial drop in occurrence may not have been observed because macrolides, such as tylosin, were still being administered to swine for therapeutic purposes (Aarestrup et al. 2000). In the same study, similar drops in occurrence were observed for erythromycin- and virginiamycin-resistant *Enterococcus* spp. isolates in broilers, and for glycopeptides-resistant *E. faecium* isolates in swine (Aarestrup et al. 2000). These findings were further confirmed by similar research conducted by Emborg et al. (2003) in Denmark on the occurrence of antimicrobial resistant *Enterococcus* spp. in broilers. One of the ways in which resistant pathogens can be transferred to humans is via the consumption of meat products, which is beyond the scope of this report. The National Antimicrobial Resistance Monitoring System (NARMS), a collaboration between the USFDA, the USDA, and the Centers for Disease Control and Prevention (CDC), conducts annual surveys of the prevalence of resistant pathogens on meat products (see NARMS, 2009) and provides further information.

Research indicates a higher prevalence of antimicrobial-resistant strains of pathogens in livestock and poultry handlers compared to the general public (Swartz 2002). Levy et al. (1976) found that after tetracycline-supplemented feed was introduced on a poultry farm, tetracycline-resistant *E. coli* isolates increased in fecal samples from both the poultry and farm family members. After introducing the medicated feed, 80% of the isolates in the family members were tetracycline-resistant, compared to only 7% of isolates from neighbors. The percent of resistant isolates found in the family members decreased to levels closer to the percent detected in neighbors approximately six months after discontinuing the use of tetracycline in the animal feed. Similar findings were reported by van den Bogaard et al. (2002), who found significant correlations between the prevalence of antimicrobial-resistant *Enterococcus* spp. in broilers and broiler farmers and also between broilers and poultry slaughterers.

### 6.3.3. U.S. and International Responses to Livestock Antimicrobial Use

Making the direct link between livestock and poultry antimicrobial use and resistant infections in humans is challenging and controversial, in part because bacteria can develop resistance naturally or from antimicrobial use in humans (Levy and Marshall 2004). However, in specific cases, years of research and evidence have demonstrated the link between livestock and poultry antimicrobial use and resistant infections in humans, leading to limitations or bans on certain antimicrobials. Most recently, because of the relationship between livestock and poultry antimicrobial use and the evolution and proliferation of antimicrobial-resistant pathogens, a federal court ordered the USFDA to evaluate the human health risks associated with livestock and poultry antimicrobial use (see Federal Court Ruling text box). The USFDA also recently banned the use of cephalosporin in livestock and poultry, related to

---

**Federal Court Ruling Requires USFDA to Evaluate Human Health Risks Associated with Livestock Antimicrobial Use**

Recent federal court decisions ordered the USFDA to re-evaluate the human health implications of the use of antimicrobials in livestock feed. The U.S. District Court for the Southern District of New York rulings came in response to a suit brought by the Natural Resources Defense Council, the Union of Concerned Scientists, and others. In a March, 2012 ruling, which USFDA is currently appealing, the federal judge required USFDA to withdraw its approval for most non-therapeutic uses of tetracyclines and penicillin in livestock feed, unless the practices are proven to be safe for humans. Following the court order, USFDA called for drug manufacturers to voluntarily place restrictions on the use of certain drugs in livestock feed. The most recent ruling, in June, 2012, requires USFDA to withdraw its approval of the use of antimicrobials in livestock unless industry can prove they are safe. (References: Jacobs 2012,

---

antimicrobial resistance (see Cephalosporin text box). In 2005, the USFDA banned the use of fluoroquinolone in the poultry industry because substantial data and research indicated that an increase in human infections caused by fluoroquinolone-resistant *Campylobacter* was associated with poultry consumption (Nelson et al. 2007). The fluoroquinolone ban is anticipated to reduce the selective pressure not only on fluoroquinolone-resistant *Campylobacter* but also on non-typhodial *Salmonella* species and other foodborne pathogens that can cause infections in humans (Nelson et al. 2007).

In other countries, bans on the use of certain antimicrobials in livestock and poultry related to human health concerns have been in effect for decades. The sub-therapeutic use of antimicrobials in food animals has been banned in Sweden since 1986 and in Denmark since 1998 (Emborg et al. 2003, PCIFAP 2008). In 2006, the European Union banned the use of all growth-promoting antimicrobials after having already previously banned the use of human medicines from being added to livestock feed (Europa 2005). Studies conducted by Aarestrup (2000) and Emborg et al. (2003) suggest that, as a result of these bans, there have been demonstrated reductions in the occurrence of antimicrobial-resistant pathogens in livestock and poultry. However, the European Union still considers the prevalence of antimicrobial resistance a growing health problem. In November 2011, it published the Action Plan on Antimicrobial Resistance, which, among other goals, calls on European Union countries to ensure that antimicrobials are only available via prescription and to better track cases of resistance (Europa 2011).

### 6.3.4.  Summary and Discussion

Livestock and poultry antimicrobial use in the U.S. is an estimated four times greater than the amount used to treat human infections (Loglisci 2010). Research conducted by the USGAO, the WHO, and others demonstrate that overuse and misuse of antimicrobials – in humans and/or livestock and poultry – may contribute to the prevalence of antimicrobial resistance (WHO 2000, Levy and Marshall 2004, USGAO 2011a). Research has demonstrated an increased prevalence of antimicrobial-resistant bacteria on and near livestock and poultry production facilities related to the use of antimicrobials (Hayes et al. 2004, Kumar et al. 2005, Sapkota et al. 2011). Antimicrobial-resistant pathogens have been detected in meat products (NARMS 2009). What is less clear is the extent to which antimicrobial-resistant human infections are related to the use of antimicrobials in livestock and poultry. Making that connection is challenging – USFDA reviewed decades of scientific research before banning fluoroquinolone use in poultry in 2005 and prohibiting prophylactic use of cephalosporin in certain types of livestock in 2012 (Nelson et al. 2007, USFDA 2012b).

As noted by Kumar et al. (2005), significant costs incur when antimicrobials used to treat human, pet and/or livestock and poultry bacterial infections become ineffective because of resistant bacteria. These costs are related to increased health costs and loss of livestock and poultry, as well as the need to develop new drugs. More representative data about the occurrence of antimicrobial resistance in different types of livestock and food products will help researchers and agencies identify trends and better understand the relationships between livestock and poultry antimicrobial use, the prevalence of resistant pathogens, and the occurrence of human infections caused by resistant pathogens.

## 6.4. Endocrine Disruption

Livestock excrete natural hormones (i.e., estrogens, androgens, and progestogens), and synthetic hormones (i.e., trenbolone acetate, zeranol, and melengestrol acetate in the case of some cattle). These hormones can enter aquatic ecosystems through runoff following manure land application, wash-off from farming operations, or via spills, overflow, and leaks from manure lagoons (Pérez and Barceló 2008). To regulate metabolic and developmental processes in animals, hormones are naturally biologically active at very low concentrations (ng/L). Even low levels of hormones detected in surface water have been implicated in endocrine disruption, adversely impacting the reproductive biology, physiology, and fitness of fish and other aquatic organisms (Zhao et al. 2008). To date, the majority of research has been conducted on the environmental impacts of hormones from human waste streams (e.g., municipal wastewater treatment plant discharges). However, recent research suggests that exposure to animal manure can also have endocrine-disrupting effects on aquatic organisms (Lee et al. 2007, Ciparis et al. 2012).

> ✓ Hormones are endocrine system regulators that are biologically active even at low concentrations.
>
> ✓ Fish exposure to estrogens can cause defeminization in females and demasculinization in males, reducing reproductive fitness.
>
> ✓ The biological activity of the synthetic hormone melengestrol acetate is estimated to be nearly 125 times greater than that of natural progesterone.

Sex steroids regulate the differentiation and structural development, as well as behavior and function, of the reproductive system in vertebrates (Lange et al. 2002). Specifically, estrogens are responsible for the development and maintenance of female sex organs and characteristics, while androgens are responsible for male organs and characteristics. Progestogens are involved in the female menstrual cycle and pregnancy. An investigation into the ecological toxicity of 92 types of hormones using USEPA's ECOSAR program found that hormones exhibited the greatest toxicity to aquatic biota, compared to several other classes of pharmaceuticals (Sanderson et al. 2004). The study predicted that 80% of the compounds were very toxic and 52% extremely toxic to fish based on impacts on species survival and reproduction. The study found that only 1% of hormone compounds were non-toxic to fish, daphnids, or algae, illustrating the potential ecological effects associated with hormones in surface waters.

The majority of research on hormones in surface waters has been conducted on estrogens, which can cause physiochemical changes in sensitive fish and other aquatic organisms. Fish exposure to exogenous estrogens can induce the production of egg yolk precursor proteins (vitellogenin) and eggshell proteins (zona radiata), which are associated with reduced testicular growth, reduced testicular and ovary size, decreased egg production, and liver and kidney damage (Lange et al. 2002). Exposure to exogenous estrogen can also lead to reduced reproductive fitness, intersex (the presence of both male and female sex characteristics), skewed population sex ratios, abnormal spawning behavior, and compromised immune systems in fish (Iwanowicz and Blazer 2011). The most potent estrogen metabolite is 17β-estradiol, which has been associated with adverse impacts on gamete production, maturation, spawning, and sexual differentiation in a variety of fish species (Lange et al. 2002, Zhao et al. 2008).

Exposing fish to animal manure containing natural hormones has also been shown to cause adverse impacts on fish, though research on hormones in manure is limited at this time (the majority of research is focused on aquatic life impacts from hormones in wastewater treatment plant discharges). Orlando et al. (2004) found that exposure of wild fathead minnows to animal feedlot effluent caused defeminization in females and demasculinization in males (i.e., reduced testicular size and testosterone synthesis, and altered head morphometrics). As suggested by the author, results from this study indicate that there were potent androgens and estrogens in the feedlot effluent. A separate study reported a high intersex prevalence in male smallmouth bass in the Potomac River Basin in the Mid-Atlantic region. This was partly explained by hormone contributions from runoff containing livestock (primarily poultry) manure within the watershed (Blazer et al. 2007).

Exposure to synthetic hormones and their metabolites from livestock and poultry manure can also adversely impact the reproductive endocrinology of some fish. Fathead minnow fecundity can be reduced when exposed to 17β-trenbolone and 17α-trenbolone (metabolites of trenbolone acetate) at concentrations greater than 27 ng/L, and 16 ng/L for 21 days, respectively (Ankley et al. 2003, Jensen et al. 2006). For perspective, concentrations of 17β-trenbolone have been detected in runoff from beef cattle feedlots at concentrations of up to 20 ng/L, which is slightly lower than the documented levels of concern (Durhan et al. 2006). However, 17α-trenbolone has been documented at concentrations ranging from <10 to 120 ng/L, which are high enough levels to potentially have adverse impacts (Durhan et al. 2006). Importantly, this information is based on a limited number of studies, and further research is needed to truly understand whether levels observed in surface waters are sufficient to cause adverse effects on aquatic life.

The hormone 17β-trenbolone is considered a potent androgen because it binds with greater affinity to the androgen receptor of fathead minnows than naturally-produced testosterone (Ankley et al. 2003). Research conducted by Jensen et al. (2006) suggests that 17α-trenbolone may be just as potent as 17β-trenbolone. Exposure to the trenbolone acetate metabolites can also result in the formation of dorsal (nuptial) turbercles on females: these tubercles are normally present on spawning males (Ankley et al. 2003, Jensen et al. 2006). In another study, male fathead minnows exposed to fecal slurry from cattle implanted with trenbolone acetate and estradiol experienced demasculinizing and feminizing effects (Sellin et al. 2009). Currently, there are no published studies on the potential adverse impacts of synthetic progestins on aquatic organisms. However, Schiffer et al. (2001) and Lee et al. (2007) provide evidence suggesting that the progestinal activity of melengestrol acetate is estimated to be nearly 125 times greater than that of progesterone.

The presence of hormones in aquatic ecosystems is not new since all mammals naturally produce and excrete hormones. In the past decade, a number of studies, most of which have been focused downstream from wastewater treatment plant discharges, have suggested potential adverse impacts of hormones on the endocrinology of fish (Lee et al. 2007). Additionally, a limited number of case studies suggest that hormones from manure specifically, may have similar endocrine-disrupting impacts on aquatic life (i.e., Blazer et al. 2007). Little is known about the potential adverse impacts of long-term exposure to hormone doses lower than those exhibiting a response over a 21 day test, such as in the previously discussed studies conducted by Ankley et al. (2003) and Jensen et al. (2006). Importantly, the detection of hormones in the environment is relatively new because recent advancements in laboratory methods and analytical techniques have made it possible to detect hormones, which are often present in low concentrations (ng/L) in the environment (Lee et al. 2007). The ability to detect hormones in the environment has allowed for more research on the potential impacts of hormones from human and animal waste streams on aquatic organisms. Given the adverse impacts of exogenous hormones on aquatic organisms, the increasing amount of both natural and synthetic hormones entering the environment through livestock animal manure needs additional review, particularly because some synthetic hormones (e.g., trenbolone acetate) appear to be more stable in the environment than natural hormones (Ankley et al. 2003, Lee et al. 2007).

## 6.5. Waterborne Disease Outbreaks

Livestock and poultry manure can contain pathogens with zoonotic potential (transferred to humans from other animals) (e.g., Rogers and Haines 2005). Land application of manure presents opportunities for those pathogens to enter recreational waters and drinking water sources, potentially leading to a waterborne disease outbreak (see Chapter 3). Exposure of crops to manure or contaminated water can also lead to foodborne illness.

Although the majority of waterborne disease outbreaks have been attributed to human fecal contamination (Rosen 2000), investigations have identified pathogens in manure as a possible or confirmed source in a number of outbreaks (Rosen 2000, Guan and Holley 2003). A number of examples of outbreaks are briefly described in Table 6-3, which also includes outbreaks caused by contamination of food with manure. This chapter reviews waterborne disease outbreaks, presents examples of notable outbreaks, and notes informational gaps, particularly in the ability to trace the origin of waterborne diseases in many cases.

**Table 6-3. Waterborne and foodborne disease outbreaks.** *(Table 6-3 continues on the following page.)*

| Location | Year | Pathogen | Suspected Source of Contamination | Predominant Illness and Impact | References |
|---|---|---|---|---|---|
| Nova Scotia, Canada | 1981 | *Listeria monocytogenes* | Cabbages grown on a farm fertilized with *Listeria*-contaminated sheep manure. | 41 cases of listeriosis, 18 deaths | Health Canada 2009 |
| Carrollton, GA | 1987 | *Cryptosporidium parvum* | Runoff from cattle grazing areas and a sewage overflow-contaminated river water used for drinking water supply. Also, drinking water treatment deficiencies. | 13,000 cases of cryptosporidiosis | Solo-Gabriele et al. 1996, USEPA 2004a |
| Ayrshire, UK | 1988 | *Cryptosporidium parvum* | Post-treatment contamination of a municipal drinking water tank with runoff; cattle manure slurry sprayed nearby. | 27 confirmed cases, hundreds more suspected | Smith et al. 1989 |
| Swindon & Oxfordshire, UK | 1989 | *Cryptosporidium parvum* | Oocysts in runoff from fields with cattle entered water supply (Thames River) after heavy rains. | 516 cases of cryptosporidiosis over 5 months, mostly children, 8% hospitalized | Richardson et al. 1991, USEPA 2004a |
| Cabool, MO | 1990 | *E. coli* O157:H7 | Contamination of distribution system with human sewage overflow via water main breaks and meter replacements. Community practices dairy farming. | 243 cases of diarrhea, including 86 with bloody diarrhea, 32 hospitalized, 2 Hemolytic-uremic syndrome (HUS), 4 deaths | Geldreich et al. 1992, Swerdlow et al. 1992, Cotruvo et al. 2004 |
| Bradford, UK | 1992 | *Cryptosporidium parvum* | *Cryptosporidium* oocysts in the water supply after heavy rains in the catchment area. Also, deficiencies in drinking water treatment. | 125 cases of cryptosporidiosis | Atherton et al. 1995, USEPA 2004a |

| Location | Year | Pathogen | Suspected Source of Contamination | Predominant Illness and Impact | References |
|---|---|---|---|---|---|
| Maine | 1992 | *E. coli* O157:H7 | Cow manure spread in a vegetable garden. | 4 cases of bloody diarrhea, one adult and 3 children | Cieslak et al. 1993, USEPA 2004a |
| The Netherlands | 1993 | *E. coli* 0157:H7 | Illness was contracted swimming in a semi-natural shallow lake. Possible sources include human excrement and water from ditches draining meadows with cattle. | 12 cases of enteritis, 5 children with HUS | Cransberg et al. 1996, Cotruvo et al. 2004 |
| Milwaukee, Wi | 1993 | *Cryptosporidium parvum* | *Cryptosporidium* oocysts in drinking water source, related to heavy rain and increased turbidity. Source may have been animal manure and /or human excrement. | 403,000 cases of cryptosporidiosis, 54 deaths | MacKenzie et al. 1994, Hoxie et al. 1997 |
| Sakai City, Japan | 1995 | *E. coli* O157:H7 | Animal manure used in fields growing alfalfa sprouts. | 12,680 cases among schoolchildren, most with diarrhea or bloody diarrhea. 121 cases of HUS, 425 hospitalized, 3 deaths | Fukushima et al. 1999, USEPA 2004a, Rogers and Haines 2005 |
| Connecticut and Illinois, USA | 1996 | *E. coli* O157:H7 | Consumption of mesclun lettuce. Cattle were found near the lettuce fields. | 53 cases, 40 with bloody diarrhea, and 3 HUS cases | Hilborn et al. 1999 |
| Washington Co., NY | 1999 | *E. coli* O157:H7 and *Campylobacter* spp. | Contamination of un-chlorinated water supply well used by food vendors for ice and drinks. Possible sources are of cattle or human origin. | Bopp et al. cite 775 cases, 65 hospitalized, 11 HUS cases, 2 deaths

CDC cites 921 persons with diarrhea after attending fair | CDC 1999, Bopp et al. 2003, Cotruvo et al. 2004 |
| California, USA | 1999 | *E. coli* 0157:NM | Recreational exposure to lake water; fecal contamination may have been from humans, cattle, or deer. | 7 cases of diarrhea in children | Feldman et al. 2002, Cotruvo et al. 2004 |
| Walkerton, Canada | 2000 | *E. coli* O157:H7 and *Campylobacter* spp. | Runoff from farm fields entering a shallow well used for the town's water supply. | 2,300 cases of diarrhea, more than 100 hospitalized, 27 HUS cases, 6 deaths | Valcour et al. 2002, Hrudey et al. 2003, Cotruvo et al. 2004, USEPA 2004a, PHAC 2000 |
| Cornwall, U.K. | 2004 | *E. coli* 0157:H7 | Exposure to a freshwater stream crossing a seaside beach; the stream had cattle grazing upstream. | 7 cases in children, diarrhea and bloody diarrhea, 4 hospitalized | Ihekweazu et al. 2006 |

### 6.5.1. Routes of Exposure and Example Outbreaks

A waterborne disease outbreak is defined by two criteria: 1) two or more persons experience an illness and are linked epidemiologically by time, location of exposure to water, and illness characteristics, and 2) the epidemiological evidence implicates water as the source of illness (Hlavsa et al. 2011). Humans may be exposed to waterborne pathogens via contact with treated or untreated recreational water or ingestion of drinking water (Bowman 2009). Although exposure may also occur through inhalation of some organisms (e.g., *Legionella pneumophila, Naegleria fowleri, Acanthamoeba*), this method of exposure is outside of the scope of this report and is not discussed further. Surface waters may become contaminated by zoonotic pathogens from agricultural or urban runoff, although dilution and die-off can help mitigate the possibility of illness (Rosen 2000). Ground water may become contaminated through infiltration of agricultural runoff or leaching of land-applied manure (Marks et al. 2001), with shallow aquifers and fractured rock and karst aquifers being especially vulnerable. Agricultural or urban runoff may also enter inadequately protected private or municipal wells (Rosen 2000).

Large and/or intense precipitation events can increase the likelihood of contamination of water with microorganisms carried in runoff and/or through impacts on drinking water treatment processes. Such hydrologic conditions in an agricultural watershed raise the possibility of waterborne disease outbreak due to zoonotic organisms in manure. Curriero et al. (2001) analyzed the relationship between precipitation and waterborne disease based on all reported waterborne disease outbreaks in the U.S. from 1948 to 1994. Of 548 waterborne disease outbreaks analyzed, 51% were observed to coincide with extreme precipitation events. A number of examples can be found in which a combination of heavy rainfall and deficient treatment of a surface water supply resulted in a waterborne disease outbreak; some were outbreaks in which manure was a suspected source. For example, insufficient chlorination related to increased turbidity from heavy precipitation was implicated in a 1978 *Campylobacter* outbreak in Bennington, Vermont, with 3,000 cases (Vogt et al. 1982). In this outbreak, the main water source for the town was vulnerable to deficient sewer systems as well as animal excrement on the banks (animal type unknown); increased runoff from the watershed provided contamination, and the additional turbidity decreased the effectiveness of the disinfection.

> ✓ Many waterborne disease outbreaks are undetected or unreported.
>
> ✓ From 1991-2002, the pathogens for almost 40% of gastrointestinal illness outbreaks associated with drinking water were not identified.
>
> ✓ Many if not most outbreaks for which the pathogen is known are attributable to human sources of infection.
>
> ✓ The number of manure-related outbreaks is not known, but contamination from manure has been suggested as a possible causative agent in a number of outbreaks involving zoonotic pathogens.

The Milwaukee outbreak (March and April, 1993) was the largest drinking water-related *Cryptosporidium* outbreak on record and was related to heavy precipitation and drinking water treatment deficiencies. An estimated 403,000 people were affected, and 54 deaths were reported (Hoxie et al. 1997). Milwaukee uses water from Lake Michigan and has two treatment plants; the locations of cases of illness suggested that one of the two plants (Howard Avenue) was responsible (USEPA 2004, Bowman 2009). It is believed that heavy rainfall and snow runoff may have transported *Cryptosporidium* oocysts to Lake Michigan in addition to causing high turbidity (Rosen 2000). Plant operators may not have used adequate coagulant to treat the water (MacKenzie et al. 1994, Bowman 2009). Also, the plant recycled its filter backwash water, possibly concentrating oocysts in the plant. At the time of the outbreak, the plant met all drinking water quality standards (MacKenzie et al. 1994, Rosen 2000), but the treatment processes were not adequate to remove or inactivate *Cryptosporidium* oocysts. After the outbreak, the intake was moved and the plant was upgraded to prevent future *Cryptosporidium* outbreaks by the addition of ozone for disinfection and enhanced filter beds with continuous turbidity meters (MacKenzie et al. 1994, Bowman 2009). Also, the practice of recycling filter backwash water has been discontinued (MacKenzie et al. 1994). Possible sources of the *Cryptosporidium* include cattle manure in the watershed, slaughterhouse waste, and sewage overflow (MacKenzie et al. 1994). Genetic testing has implicated human sewage, but the analysis was based on only four isolates and may not be representative of the entire outbreak (Peng et al. 1997). Thus, the sources of the oocysts remain unclear.

Contamination of ground water supplies has also resulted in waterborne disease. In August of 1999, a large outbreak of *E. coli* O157:H7 and *Campylobacter jejuni* occurred in association with the Washington County Fair in New York State. According to the CDC (1999), 921 individuals reported diarrhea after attending the fair. *E. coli* O157:H7 was cultured from stools from 116 persons, with 13 also infected with *Campylobacter*. Two deaths were reported. Water at the fairgrounds was supplied by six shallow wells, four of which were un-chlorinated (Bopp et al. 2003). One of the un-chlorinated wells was implicated in the outbreak. Two possible sources of contamination were located near the well: a cow manure storage site and a dormitory septic tank. The well may have been contaminated by runoff resulting from a heavy rainfall that occurred during one day of the fair.

An *E. coli* O157:H7 outbreak linked to cattle manure contamination of a ground water supply occurred in May 2000 in Walkerton, Ontario, resulting in more than 2,000 cases. Of those, 27 people developed hemolytic-uremic syndrome (HUS), and there were six deaths. Both *E. coli* O157:H7 and *Campylobacter* were confirmed in stool samples from those infected (PHAC 2000). Testing of one of the town's production wells and the distribution system demonstrated evidence of fecal contamination of the drinking water, and DNA analyses by polymerase chain reaction (PCR) confirmed the presence of *E. coli* O157:H7 (PHAC 2000). To determine the origin of the *E. coli* O157:H7, 13 livestock farms were investigated in the area. *Campylobacter* was found on nine farms, and both *E. coli* O157:H7 and *Campylobacter* were found on two farms, including a farm near the tested drinking water well (PHAC 2000). Typing of isolates, including the use of genetic fingerprinting, matched the isolates from the farm near the well to those found in most of the patients (PHAC 2000, Clark et al. 2003). The analysis indicates that the outbreak was caused by a combination of factors including flooding from heavy rainfall, runoff contaminated by cattle manure, a well vulnerable to surface water contamination (as further indicated by historic records), and decreased disinfection efficacy due to increased turbidity (PHAC 2000, Clark et al. 2003).

Contamination can also occur post-treatment, as was the case with a *Cryptosporidium* outbreak in Ayrshire, England in 1988. Twenty-seven cases of cryptosporidiosis were confirmed, although inquiries by local health authorities suggested that there may have been hundreds of cases. The contamination was traced to intermittent seepage of runoff into a clay pipe that fed into a water tank. Cattle manure slurry had been sprayed nearby, and there had been heavy rain, which would have increased water leakage into the tank (Smith et al. 1989).

If contaminated irrigation water or runoff reaches crops or if manure is applied to fields, foodborne outbreaks may also occur; two thirds of deaths from food-borne outbreaks are attributed to zoonotic bacterial pathogens: *Salmonella* sp., *Listeria monocytogenes*, *Campylobacter*, and *E. coli* O157:H7 (Bowman 2009). A variety of fresh fruits, vegetables, and nuts may be affected (Rogers and Haines 2005, CDC 2013).

### 6.5.2. Outbreak Statistics

Data on waterborne disease outbreaks in the U.S. are compiled and reported by the CDC, the Council of State and Territorial Epidemiologists, and the USEPA through the Waterborne Disease and Outbreak Surveillance System (WBDOSS), a voluntary system in place since 1978. Reports are published by the CDC as surveillance summaries, allowing for an assessment of trends in the prevalence of different types of pathogens in recreational and drinking waters. Although these reports do not identify potential animal vs. human sources for outbreaks, they do provide information on the types of illness and the etiologic agents, some of which can be zoonotic. These reports, however, are recognized as underestimates of the true number of outbreaks because of unreported or unrecognized cases (see subsection 6.5.3).

During 2007 and 2008, 36 drinking water-related disease outbreaks were reported to the CDC (Hlavsa et al. 2011); 12 were related to untreated ground water used for drinking, and seven were attributed to treatment failures; these 19 outbreaks all resulted in acute gastrointestinal illness. For recreational water, 134 outbreaks causing nearly 14,000 cases of illness were reported in the same time period (Hlavsa et al. 2011). Outbreaks of acute gastrointestinal illness can be caused by pathogens with zoonotic potential (Rosen 2000). For example, among 21 bacterial outbreaks associated with drinking water during 2007-2008, four were caused by *Campylobacter*, three by *Salmonella* (including one outbreak with 1,300 cases), and one by *E. coli* O157:H7. (Other bacterial outbreaks were caused by *Legionella pneumophila*, which is not considered zoonotic). Two of the three parasitic outbreaks were caused by *Giardia intestinalis* (synonymous with *Giardia lamblia*). Norovirus was responsible for four of the five viral outbreaks. Among 134 recreational water disease outbreaks in 2007-2008, *Cryptosporidium* caused 60 outbreaks, most of which were caused by exposure to treated water such as chlorinated swimming pools and spas (Hlavsa et al. 2011).

### 6.5.3. Limitations Associated with Detection of Zoonotic Waterborne Disease Outbreaks

Determining the pathogen and tracing the origin of a waterborne disease outbreak can be challenging. Therefore, the causes of outbreaks often remain unknown, including those that may be related to livestock and poultry operations. Between 1991 and 2000, for example, the pathogens associated with nearly 40% of drinking water outbreaks were not identified (Craun et al. 2006). Without knowing which pathogen is responsible for the outbreak, it is even more difficult to trace the pollution source. Livestock and poultry manure is a source of pathogens, but because of the limitations associated with tracing an outbreak back to the source, manure-related outbreaks may be left undetected or attributed to another source incorrectly or by default. For example, if an outbreak cannot be traced to water or if the route of transmission is unclear, the source may be attributed to food (Bowman et al. 2009). It is also generally recognized that reported outbreaks represent only a small portion of total outbreaks (Craun et al. 2006); more research as well as better monitoring and surveillance are needed to better understand the possible extent of underestimation.

Several factors affect whether an outbreak is recognized. Not all infected patients seek medical attention, making the number of cases difficult to track. The local health department needs to have adequate resources for surveillance and investigation (Craun et al. 2006). Also, many outbreaks may simply be too small to notice. Importantly, by the time an outbreak is discovered, the contamination may have already flushed through the water source, making it difficult to conclusively link the outbreak to water or identify the source of pollution (e.g., Hunter et al. 2003, Perdek et al. 2003). Pathogen detection methods also present challenges in terms of time requirements, method sensitivities, the abilities of the pathogens to grow in culture, and indications of viability (Perdek et al. 2003, Cotruvo et al. 2004, Yu and Bruno 1996, Pyle et al. 1999, Hunter et al. 2003, Perdek et al. 2003). These factors compound the difficulty in assessing to what degree (and where) waterborne illnesses may be caused by zoonotic pathogens transported in manure. A number of serotyping methods and molecular methods, however, may be used to attempt to determine the source of a pathogen (e.g., Hunter et al. 2003). An example of a useful development has been the identification of *Cryptosporidium* genotypes that can help determine if the source is zoonotic (e.g., Royer et al. 2002).

### 6.5.4. Summary and Discussion

Waterborne disease outbreaks can occur from exposure to contaminated recreational water or ingestion of contaminated drinking water. Although many, if not most, outbreaks are believed to be associated with human fecal contamination, livestock and poultry manure contains pathogens that may contaminate water. The number of waterborne disease outbreaks that may be associated with zoonotic pathogens from livestock and poultry manure is not understood. This is in part because confirming the source of an outbreak is challenging, and many outbreaks may not even be recognized. Not all persons will seek medical attention, some outbreaks may be too small to be noticed, and reporting to the WBDOSS is voluntary. Furthermore, among recognized outbreaks of acute gastrointestinal illness, the causative agent remains unidentified for a substantial portion (Craun et al. 2006, Hlavsa et al. 2011).

Routes of exposure to waterborne pathogens may involve entry of pathogen-contaminated water into drinking water supplies, either via runoff or infiltration, or into recreational water via runoff. Heavy rainfall in particular has been implicated in a number of outbreaks; the possibility of manure-related contamination may be greater if manure has been recently applied, allowing runoff contaminated with manure to reach recreational waters or drinking water supplies.

Agricultural sources such as runoff containing manure have been suspected in a number of waterborne outbreaks caused by pathogens with zoonotic potential (Table 6-3). It is not generally possible to confirm unequivocally that the source is agricultural as opposed to human, but watershed characteristics, such as nearby livestock and poultry operations and their proximity to recreational or drinking water resources suggest possible zoonotic transmission. Greater surveillance is needed to understand the degree to which manure-related pathogens may be implicated in waterborne disease outbreaks.

## 6.6. Potential Manure-Related Impacts Summary and Discussion

Livestock production has become increasingly concentrated in the U.S., which in turn has resulted in greater volumes of manure and associated contaminants in local areas (MacDonald and McBride 2009). This chapter reviews some of the potential and documented impacts associated with emerging contaminants, including antimicrobials and hormones. To a lesser extent, this chapter reviews pathogens and indirect effects of nutrients, which have been reviewed in detail elsewhere (e.g. Rogers and Haines 2005, Camargo and Alonso 2006, NITG 2009). The research provided in the preceding chapters indicates both documented and potential ecological and human health impacts associated with livestock and poultry manure, though overall impacts are largely unknown. Importantly, research indicates that manure runoff can contribute to water quality degradation, and the magnitude of manure generated (1.1 billion tons in 2007) may be of concern.

Aquatic communities can be adversely impacted by manure runoff or discharges to surface waters in a number of ways. Nutrient loading is the typical impact discussed, though large manure spills have been implicated in fish kills and degraded water quality (Mulla et al. 1999). Manure can also be a source of hormones, which are known endocrine disruptors. While research is limited, exposure to hormones from livestock and poultry manure has been implicated in adverse impacts on reproduction, fitness, and behavior in fish (Zhao et al. 2008, Iwanowicz and Blazer 2011).

Manure contamination of drinking and recreational water resources can be a human health concern and/or incur increased drinking water treatment costs. Nutrient loadings to surface waters may also contribute to the growth of HABs, which can produce toxins that can be harmful to human and ecological health (Lopez et al. 2008). Waterborne disease outbreaks have been associated with pathogen contributions from manure, though source detection is challenging (Rosen 2000, Guan and Holley 2003). The human health impacts related to potential long-term exposure via drinking water to low levels of hormones and antimicrobials (from all sources) are unknown. Furthermore, little is known about the potential synergistic effects between antimicrobials and hormones, which may be present in drinking water systems (Weinberg et al. 2008).

A topic of increasing interest has been the issue of widespread antimicrobial use in livestock and poultry. Such widespread use may select for antimicrobial-resistant bacteria (Swartz 2002). Many antimicrobials are also used in human clinical medicine (Sapkota et al. 2007). Related to antimicrobial resistance and human health concerns, the USFDA has banned the use of certain types of antimicrobials for livestock and poultry use (Nelson et al. 2007, Gilbert 2012).

Research pertaining to the human health and ecological impacts associated with livestock and poultry manure is relatively limited, particularly in terms of antimicrobials and hormones. However, as reviewed in this chapter, these contaminants have been detected in manure and environments proximal to livestock and poultry operations. A more thorough understanding of livestock and poultry antimicrobial and hormone use and excretion and better source tracking of waterborne disease outbreaks is needed to fully address the ecological and human health impacts associated with manure generation.

# 7. Drinking Water Treatment Techniques for Agricultural Manure Contaminants

Drinking water resources may be contaminated with livestock and poultry manure through overland runoff, soil infiltration, direct discharges or atmospheric deposition. Key manure contaminants reviewed in this report include pathogens, antimicrobials, hormones, and nutrients, though Table 1-1 provides a more complete list. Because of their acute negative human health impacts, much research and regulatory attention has been given to ensuring the removal and/or inactivation of pathogens and nutrients such as nitrate and nitrite. For example, MCLs and treatment technique requirements have been established under USEPA's Safe Drinking Water Act, focusing on the removal or inactivation of pathogens from drinking water sources (see USEPA's current drinking water regulations website: http://water.epa.gov/lawsregs/rulesregs/sdwa/currentregulations.cfm). While extensive research has been conducted on pathogens, emerging contaminants, such as hormones and antimicrobials, have only recently been studied. This is largely because of recent developments in analytical techniques that allow for the detection of such contaminants at low levels (e.g., ng/L). Research is limited, though hormones and antimicrobials have been detected in drinking water supplies (Stackelberg et al. 2007, Benotti et al. 2009), and understanding how effectively these compounds are removed by drinking water treatment processes is important for preventing potential long-term public health impacts (Snyder et al. 2008, Weinberg et al. 2008). Ingestion of antimicrobials and hormones via drinking water is likely low over the course of a lifetime, though short- and long-term effects related to low-level exposure or synergisms between different compounds are not fully understood (Weinberg et al. 2008).

This chapter provides a brief overview of watershed management techniques and drinking water treatment processes that can help to reduce surface water pollution and remove contaminants. Importantly, this chapter focuses primarily on antimicrobial and hormone removal from drinking water, because our understanding of removal of these contaminants from drinking water is relatively new given recent advancements in analytical techniques allowing for measurement of these compounds. Information on the removal of pathogens and nutrients is covered briefly, but is well established and available from other sources (USEPA's *Alternative Disinfectants and Oxidants Guidance Manual* (1999), AWWA's *Removal of Emerging Waterborne Pathogens* (2001), USEPA's *Effect of Treatment on Nutrient Availability* (2007).

## 7.1. Source Water Protection

A multi-barrier approach including source water protection efforts in addition to drinking water treatment can help minimize exposure to animal manure contaminants. The first step in this approach is to utilize source water contamination prevention measures related to livestock and poultry manure that can improve water quality and reduce the burden on drinking water treatment utilities. Management strategies include preventing animals and their manure from coming into contact with runoff and water sources; properly applying manure as fertilizer on crop or pastures during growing seasons to match crop nutrient needs (based on well-developed Nutrient Management Plans), and appropriately managing pastures (USEPA 2001).

A variety of intervention practices may be employed to minimize manure contact with precipitation and runoff. Specific practices include lining and maintaining manure storage lagoons, constructing litter storage facilities, diverting precipitation and surface water away from manure, composting, and treating runoff (Armstrong et al. 2010) (see also Chapter 8 for further information). The goal of pasture management is to protect water resources from direct livestock contact and runoff from animal feeding operations. Fencing can be used to keep livestock and poultry from defecating in or near streams or wells. Additionally, providing alternative water sources and hardened stream crossings for use by livestock lessens their impact on water quality (USEPA 2001). For more information on livestock and poultry management strategies designed to protect water resources, refer to the USEPA's *Source Water Protection Practices Bulletin Managing Livestock, Poultry, and Horse Waste to Prevent Contamination of Drinking Water* (2001).

## 7.2. Drinking Water Treatment Techniques

While source water protection efforts can help to reduce the burden for contaminant removal on drinking water treatment plants, appropriate treatment processes must also be in place. Conventional drinking water treatment facilities typically incorporate: 1) coagulation and flocculation, in which dirt, colloids and other suspended particles in the water column bind to alum or other chemicals that are added to the water to form floc; 2) sedimentation, in which the coagulated particles (floc) settle to the bottom; 3) filtration, in which particles including clays, silt and organic matter are physically removed; and 4) disinfection, in which microorganisms are killed or inactivated (USEPA 2004b). In addition, treatment facilities may utilize advanced treatment options such as nanofiltration and ultrafiltration, reverse osmosis, ion exchange and carbon adsorption to remove contaminants not removed by conventional filtration (USEPA 2004b).

The following subsections provide a brief overview of pathogen and nutrient removal and a more detailed review of recent research findings on antimicrobial and hormone removal.

### 7.2.1.   Pathogen and Antimicrobial-Resistant Bacteria Removal

Coagulation and filtration processes have been demonstrated to remove bacteria, protozoa and viruses. Maximum removal of pathogens is associated with optimized coagulant dosing and production of water with a very low turbidity. Chlorine, the most common disinfectant in the U.S., is an effective bactericide and viricide. Protozoan cyst and oocysts have been found to be more resistant to chlorine disinfection, and high contact time (CT) values are required for their inactivation. *Crypstosporidium parvum* and *Giardia lamblia* are resistant to chlorine disinfection, though UV light has been found to be an appropriate disinfection alternative. For more information on pathogen removal, refer to the USEPA's *Alternative Disinfectants and Oxidants Guidance Manual* (1999) and AWWA's *Removal of Emerging Waterborne Pathogens* (2001).

The process of chlorination during drinking water treatment has been associated with an increase in antimicrobial-resistant bacteria in treated water. During testing of drinking water source, treated, and tap water, Xi et al. (2009) found that during the treatment process, there was a significant increase in the prevalence of bacteria resistant to amoxicillin, and chloramphenicol. Chlorine-induced formation of multidrug-resistant bacteria has also been documented by Armstrong (1981) and (1982). The process by which this occurs, is not entirely known, though one potential explanation is that in the presence of chlorine, the bacteria increase their expression of efflux pumps, which pump toxins and antibiotics outside of the cell (Xi et al. 2009). Further research in this area will help elucidate the impacts of chlorination on the prevalence of antimicrobial-resistant bacteria.

### 7.2.2.   Nutrient Removal

Nutrient removal in drinking water is focused on nitrate and nitrite, related to the human health impacts briefly discussed in Chapter 6. The USEPA has established a drinking water MCL for nitrite of 1 mg/L and for nitrate-nitrogen of 10 mg/L. Ion exchange, reverse osmosis, and electrodialysis have been shown to remove nitrates/nitrite concentrations to below their MCL. For more information on nitrates and nitrites, please refer to USEPA's Basic Information about Nitrate in Drinking Water, available online at http://water.epa.gov/drink/contaminants/basicinformation/nitrate.cfm. For information on other nutrients, please see USEPA's Effect of Treatment on Nutrient Availability (2007).

### 7.2.3.    Antimicrobial and Hormone Removal

Each step of the drinking water treatment process differs in its efficacy in removing antimicrobials and hormones. Generally, concentrations of antimicrobials and hormones tend to be lower in finished (i.e., treated) water than in source water, either due to degradation or removal (Stackelberg et al. 2007, Snyder et al. 2008). For example, Stackelberg et al. (2007) measured the removal of antimicrobials in a conventional drinking water treatment plant and found that, out of seven antimicrobials detected in source water, only one persisted at detectable concentrations after treatment. In that study, erythromycin, erythromycin-$H_2O$ (an erythromycin degradate), lincomycin, sulfadimethoxine, sulfamethazine, and sulfamethoxazole, all decreased from <0.1 µg/L in source water to non-detectable concentrations in finished, treated water. Sulfathiazole persisted through treatment, though maximum concentrations decreased from 0.08 µg/L in source water to 0.01 µg/L in finished water. Reporting levels for this study ranged from 0.01 µg/L to 0.1 µg/L for the aforementioned antimicrobials.

Importantly, even when treatment appears to remove nearly all of a compound from source water, those compounds are likely still present in the treated effluent, either as degradates or in concentrations below the method detection limit (Snyder et al. 2008, Weinberg et al. 2008). Furthermore, most research has focused on commonly used antimicrobials and naturally produced, rather than synthetic, hormones. Therefore, our knowledge of the amount of antimicrobials and hormones in drinking water is essentially a function of which compounds are analyzed and the analytical methods used. According to Snyder et al. (2008), no water is 'drug free' given the variety of sources of these compounds to the environment. Although some antimicrobials may be degraded during treatment, their degradates may remain biologically active, potentially having long-term public health impacts (Dodd et al. 2005, Weinberg et al. 2008). The following subsections review available research on each treatment process in terms of its effectiveness in removing antimicrobials and hormones from source water.

#### 7.2.3.1.    *Coagulation and Sedimentation*

The effectiveness of coagulation and sedimentation in antimicrobial and hormone removal appears to vary, though the processes are generally considered to be relatively ineffective in overall removal (Westerhoff et al. 2005, Stackelberg et al. 2007). Using ferric chloride as a coagulant, Stackelberg et al. (2007) reported 33% removal of sulfamethoxazole, 47% removal of erythromycin-$H_2O$, and 60% removal of acetaminophen from source water. However, in a separate study, coagulation using ferric salt or alum did not result in any statistically significant removal of carbadox, trimethoprim, or various types of sulfonamides (Adams et al. 2002). The relative ineffectiveness of coagulation and sedimentation in antimicrobial removal is not surprising because these processes remove hydrophobic compounds, and antimicrobials tend to be hydrophilic (Weinberg et al. 2008, Chee-Sanford et al. 2009).

Coagulation using alum or ferric salt appears to be even less effective in hormone removal (Westerhoff et al. 2005). Using alum, ethynlestradiol, and androstenedione were not removed in measurable amounts, and only approximately 2% of estradiol, 5% of estrone, and 6% of progesterone were removed from source water (Westerhoff et al. 2005). Using ferric salt during coagulation resulted in similar low removals.

#### 7.2.3.2.    *Filtration and Adsorption*

Nanofiltration and reverse osmosis (RO) have been shown to be effective at removing organic compounds (Snyder et al. 2008), while ion exchange is relatively ineffective in antimicrobial removal (Adams et al. 2002). The use of nanofiltration has been shown to remove as much as 80% of chlortetracycline, but only 11% to 20% of sulfonamides (Koyuncu et al. 2008). Removal of the hormones estriol, estradiol, estrone, 17α-ethinylestradiol, and testosterone through nanofiltration range from 22% to 46% (Koyuncu et al. 2008). In a separate study, Nghiem et al. (2004) also reported effective removal of estradiol, estrone, testosterone, and progesterone by nanofiltration.

Using RO, Adams et al. (2002) reported 90% removal of carbadox, trimethoprim, and sulfonamides from Mississippi River water. Currently, limited research on RO in terms of hormone and antimicrobial removal has been conducted, and despite its apparent effectiveness, RO implementation is costly and may not always be economically feasible.

The use of activated carbon appears to be effective in removing organic compounds; however, activated carbon must be regularly replaced or regenerated in order to maintain effectiveness, and the contact time and dose are also important factors in its capacity to remove compounds (Snyder et al. 2006, 2008). As much as 21% of sulfamethoxazole and 65% erythromycin-$H_2O$ may be removed through powdered activated carbon (PAC) adsorption (Westerhoff et al. 2005). The PAC dosage may be an important factor in antimicrobial removal efficacy. Using PAC doses of 10 mg/L, Adams et al. (2002) reported that antimicrobial removal ranged from 49% to 73% in Mississippi River source water, while removal rates ranged from 65% to 100% using a PAC dose of 20 mg/L. The use of PAC also appears to be effective in removing hormones from source water, with as much as 88% of testosterone, 93% of progesterone, and 94% of estradiol removed after four hours of PAC contact time (Westerhoff et al. 2005). PAC is typically only used during certain times of the year, such as during algal blooms in the late spring or summer. The use of granular activated carbon (GAC) is expected to be effective (Adams et al. 2002), though limited research has been conducted on this process in terms of antimicrobial and hormone removal.

### 7.2.3.3. *Disinfection*

Research indicates that the disinfection process is instrumental in antimicrobial and hormone removal/degradation during water treatment (Adams et al. 2002, Stackelberg et al. 2007, Snyder et al. 2008, Weinberg et al. 2008). Depending on the treatment facility, disinfection may involve the use of chlorine compounds, ozone, or UV light treatment. Chlorine disinfectants tend to react with antimicrobials such as sulfamethoxazole, trimethoprim, ciprofloxacin, and enrofloxacin, leading to their degradation, but potentially not completely eliminating their biological effect because of the formation of degradation products (Dodd et al. 2005, Weinberg et al. 2008). Disinfection through the use of sodium hypochlorite can significantly decrease the concentration of sulfathiazole in source water (Stackelberg et al. 2007). Regarding hormone removal, Snyder et al. (2008) reported higher removal rates of estrogen than testosterone and progesterone during chlorine treatment; over 20% of testosterone and progesterone were removed, while upwards of 100% of estradiol, estriol, and estrone were removed during bench-scale analyses. Although chlorination provides critical benefits in the disinfection process, it may also lead to the formation of undesirable disinfection byproducts, which can be carcinogenic. The costs and benefits of chlorination in this regard should be further evaluated.

Ozone may be more rapid and effective than chlorine compounds in organic compound removal (Weinberg et al. 2008). Adams et al. (2002) found that concentrations of antimicrobials in Mississippi River water decreased by over 95% through the use of ozone, demonstrating the effectiveness of this disinfection method. Similarly, Snyder et al. (2005) found that sulfamethoxazole concentrations in drinking water decreased from 9.7 ng/L in source water to below the detection limit (<1 ng/L) in treated water after ozonation. Ozone has also been shown to oxidize nearly 100% of testosterone, progesterone, and estrogen hormonal compounds, suggesting that ozonation is more efficient in removing hormones than is chlorination (Snyder et al. 2008). Similar results were observed by Westerhoff et al. (2005) in terms of hormone removal through the use of ozonation.

UV light alone appears to be less effective than chlorination and ozonation in removing hormones and antimicrobials (Snyder et al. 2008). Also, the dose of UV light typically used for disinfection to kill microorganisms is orders of magnitude lower than what would be required to remove micropollutants such as organic compounds (Snyder et al. 2003). However, a combination of UV light and hydrogen peroxide appears to be effective in hormone removal (Rosenfeldt and Linden 2004) and antimicrobial removal (Weinberg et al. 2008, Giri et al. 2011). Certain antimicrobials including tetracycline, chlortetracycline, and oxytetracycline may undergo photodegradation under UV light, the rate of which markedly increases when low concentrations of hydrogen peroxide are added to the disinfection process (López-Peñalver et al. 2010).

## 7.3. Summary and Discussion

Conventional drinking water treatment processes are effective at removing pathogens, and some treatment plants employ additional processes that effectively remove nutrients. Recent research indicates that conventional drinking water treatment practices are also effective in decreasing the concentrations of hormone and antimicrobials in source water, particularly during disinfection (Adams et al. 2002, Snyder et al. 2008). Filtration using nanofiltration and reverse osmosis is highly effective in antimicrobial and hormone removal (Koyuncu et al. 2008), though these processes are not always used in conventional drinking water treatment facilities, and limited research is available. Antimicrobials and hormones, as with all organic compounds, vary widely in physical and chemical characteristics and may be rapidly removed or unaffected by certain drinking water treatment processes. Therefore, antimicrobial and hormone removal from drinking water may be enhanced through the implementation of multiple treatment and disinfection methods (Snyder et al. 2008). Whereas public water systems are subject to drinking water treatment processes, private drinking water wells are typically not tested or treated for these compounds, so antimicrobials and hormones in private groundwater drinking water systems affected by livestock and poultry production may remain undetected. A stronger understanding of the prevalence and concentrations of antimicrobials and hormones in drinking water, as well as more research on which treatment processes best remove these compounds, will help in planning strategies to minimize their consumption and any potential associated health effects.

This page intentionally left blank.

# 8. Managing Manure to Control Emerging Contaminants

Historically, the focus of manure management has been on utilizing the nutrients in manure for crop production. In recent decades, livestock and poultry producers, land grant universities, and government agencies have worked together to develop practices and systems to minimize the impact of manure production and utilization on air and water quality, including drinking water. Though the practices and systems promoted by these programs typically do not focus specifically on the potential connections between manure, pathogens, emerging contaminants, and water quality, they do address many of the potential pathways described in this report (e.g., erosion, runoff, infiltration). Widespread implementation of appropriate practices and systems will help to reduce agricultural runoff and minimize the potential environmental problems associated with emerging contaminants from livestock and poultry manure.

This chapter provides a brief overview of the standard basic strategies for managing manure and a summary of additional approaches that can provide further benefits, including economic benefits. Many of the existing programs and standards described within this chapter are managed by the USDA Natural Resources Conservation Service (NRCS). Partnerships between federal agencies (including USDA and USEPA), conservation professionals, university extension offices, and local producers have formed to develop programs and technical standards that conserve natural resources, reduce soil erosion, decrease pollutant loading to the nation's surface waters, and improve source water protection. This overview is not intended to be exhaustive; the objective is to highlight information that is most relevant to individuals working to improve water quality. To learn more about tools, policies, technical standards, and programs that may not be listed here and may be more relevant to a specific location, contact your state or local NRCS District Conservationist or your area's Cooperative Extension Service. A sampling of online resources that are available to help planners and producers related to manure management are listed in Appendix 3.

## 8.1. Land Application of Manure

Manure serves as a nutrient-rich natural fertilizer and is commonly applied to cropland. In some cities, however, facilities that serve as holding pens before slaughter may discharge to wastewater treatment operations instead of land-applying the manure. Variations in the operational characteristics of livestock and poultry facilities (e.g., layout, herd size, access to forage crops and pastures, etc.) make it challenging to identify specific practices that implement widely-accepted principles regarding the timing, location, and rate of manure land application. Thus, NRCS has placed increased emphasis on meeting overarching resource conservation objectives through the development and implementation of nutrient management plans that determine the location and amount of manure applied to meet crop needs and keep manure out of surface and ground water resources. Appropriately managing manure as part of a nutrient management plan should also minimize the loading of other emerging contaminants, though there is relatively little research available that specifically addresses the consequences of manure management on emerging contaminants. In addition, there are many financial incentives to developing and implementing a nutrient management plan, including cost savings within the operation and increased access to federal financial assistance programs.

The NRCS Conservation Practice Standard 590 provides criteria for nutrient management through land application (http://www.nrcs.usda.gov/Internet/FSE_DOCUMENTS/stelprdb1046433.pdf). Producers receiving financial support from USDA for nutrient management must follow this standard.

The USEPA also requires nutrient management plans for any operation seeking a permit under the national pollutant discharge elimination system (NPDES) program. (See discussion under 8.5. CAFO Discharge Regulations). Any operation seeking NPDES permit coverage must submit a nutrient management plan as part of its permit application to be covered by an individual permit or a notice of intent to be covered by a general permit (40 CFR 122.23(h) and 122.42(e)(1)). A nutrient management plan is a manure and wastewater management tool that every permitted CAFO must use to properly manage discharges from the production or land application areas through the use of best management practices.

The regulations specify nine minimum requirements that must be included in the nutrient management plan, to the extent that they are applicable (40 CFR 122.42(e)(1)). The NPDES nutrient management practices were developed to be consistent with the content of comprehensive nutrient management plans as defined by USDA in the *Comprehensive Nutrient Management Plan Technical Guidance*. However, there are some differences between the requirements of a nutrient management plan for NPDES permitting and a comprehensive nutrient management plan as defined by USDA. The USEPA describes nutrient management planning requirements in the 2012 Technical Manual for Concentrated Animal Feeding Operations, available at http://cfpub.epa.gov/npdes/afo/info.cfm#guide_docs.

There are many resources available to assist producers with the development of nutrient management plans, including online tools (see Appendix 3) and individual consultation services provided by crop consultants, NRCS, conservation districts, and university extension personnel.

## 8.2. Manure Storage

Manure storage enables livestock and poultry producers with confined operations to better implement their nutrient management plans and apply their manure to address crop needs. Adequate storage capacity enables operators to store manure during times of the year when no crops are growing and avoid applying manure on frozen or snow-covered ground, immediately before, during, or after precipitation events, or when the land is saturated (Zhao et al. 2008). Storing manure for extended periods of time may also minimize pathogen loads and promote degradation or adsorption of antimicrobials and hormones (Shore et al. 1995, Lee et al. 2007).

Thoughtful design of manure storage infrastructure is critical for ensuring there is adequate capacity to prevent spills and over-topping of an open structure. Operational practices, such as covering open storage lagoons, are also important for preventing the addition of precipitation and managing manure volumes. The NRCS provides additional location-specific information about the design and operation of manure storage structures in their Technical Standards.

*Diverting Rainfall.* Constructing diversions and gutters around animal lots and buildings are inexpensive and effective ways to minimize the amount of water falling on and washing across manure covered areas. Diverting rainfall from areas with manure is often the first step in reducing the amount of runoff that must be managed to avoid pollution issues. The USEPA requires diversion of clean water, as appropriate, for operations with NPDES permit coverage. Clean water includes, but is not limited to, rain falling on the roofs of facilities and runoff from adjacent land.

*Storage Structures.* There are many common types of storage structures, including walled enclosures, lagoons, earthen ponds, above-ground tanks and under-floor storage pits. The size and choice of storage structure depends on multiple factors, including the animal production system, precipitation patterns, siting or design limitations, bedding materials, availability of on-site and off-site transportation options, local and state regulations, and costs. Following construction, storage structures should be checked periodically for leaks to prevent contamination of surface water and ground water. Also, insufficient storage capacity increases the risk of runoff from manure piles and spills from lagoons and other containment structures. Furthermore, it increases the possibility that an operation will have to land apply during periods of increased risk to surface water (e.g., during rainfall events).

## 8.3. Treating Manure

On some farms and in some geographic areas, the amount of manure produced from livestock and poultry operations exceeds what can be safely applied to nearby croplands or pastures to meet nutrient needs. To manage surplus manure, technologies have been developed to treat manure nutrients such that additional options for disposition of nutrients become viable. Recent research indicates that some of these technologies and processes may also promote removal and degradation of pathogens, antimicrobials, and hormones. Although many of these technologies have been proven from an engineering perspective, the costs are generally prohibitive for most producers. Livestock and poultry producers need to analyze the economic viability of any of these technologies for their specific operations. However, potential economically beneficial options do exist such as the sale of electricity generated through the manure-to-energy process. In some cases, nutrients from manure, such as phosphorus byproducts, can be recovered, sold and transported to locations low in phosphorus (Szogi et al. 2010). Given that phosphorus is a nonrenewable resource, it is anticipated that these byproducts could become an increasingly valuable source of income (Chesapeake Bay Commission 2012).

### 8.3.1. Physical and Chemical Treatments

Physical and chemical treatments are designed to separate the solids and liquids in manure slurry to make the manure easier to utilize, handle, and transport. For example, as recommended in an Ohio State University Extension manure management guide, solids may be reused for livestock bedding material, and liquids can be recycled for washing down hard surfaces (James et al. 2006).

*Physical treatment* of manure involves separating solids from liquid manure through settling, filtration, screening, or drying. Settling basins are used to separate solids through natural settling so that the solids can be removed (James et al. 2006). Solids may also be separated out in a mechanical centrifuge or through filtering and screening systems that remove solids as the liquid waste passes through. Filtering systems may be constructed with sand drying beds, stationary or vibrating screens, or vacuum filters (James et al. 2006). Manure may also be dried passively (i.e., spread in a manner that allows water to evaporate), though this method is more time consuming and is more likely to result in the emission of foul odors and greenhouse gases unless additional steps are taken to capture the emissions. The effects of physical treatment on emerging contaminants are unknown.

*Chemical treatment* involves the addition of coagulants, such as lime, alum, and organic polymers to manure (James et al. 2006). Coagulants are effective at separating solids and liquids, but the agents may persist in the manure and may reach surface waters and ground water through runoff and infiltration, if land applied. Some coagulants decrease the presence of pathogens, such as quick lime (CaO) or hydrated lime (CaOH), which increase pH and kill most microorganisms (James et al. 2006). Adding lime, however, results in an immediate loss of ammonia from the manure through volatilization (James et al. 2006), reducing its quality as a fertilizer and creating air quality concerns. The effects of chemical treatment on emerging contaminants in manure are largely unknown.

### 8.3.2. Biological Treatment Techniques

Biological treatment of manure occurs within traditional manure storage structures and other less traditional methods such as composting and anaerobic digestion. These methods remove pathogens and can reduce the total volume of manure. This subsection focuses on less traditional treatments: composting and anaerobic digestion.

### 8.3.2.1. Composting of Manure

Composting is the process of aerobic biological decomposition of manure in a controlled environment. During composting, microorganisms decompose the manure, increasing the temperature and inactivating pathogens. Numerous factors influence the effectiveness of composting, including nutrient balance (i.e., carbon to nitrogen ratio), water content, oxygen availability, porosity, and temperature (James et al. 2006). Composting manure prior to land application provides some benefits, including reduction of odor and fly problems and weed seeds (USDA 2009j). When composting is properly controlled, most pathogens are inactivated at higher temperatures (i.e., greater than 55° F), with the exception of some viruses and worm eggs (Rosen 2000, Olson 2001, Venglovsky et al. 2009). Also, the quality of the manure as a fertilizer increases when composted, because the nitrogen becomes more stable and nutrients are released more slowly than they are from raw manure (Zhao et al. 2008, USDA 2009j), though nitrogen volatilization during composting reduces the total amount of nitrogen available in the manure. When composting is used as part of a system that includes separation of liquids and solids, the practice can reduce the total amount of dry matter by 50% to 75%, with greater reductions for swine and dairy cow manure, and the total volume of manure can be reduced by as much as 85% (USDA 2007c).

Recent research suggests that composting may promote antimicrobial degradation (Zhao et al. 2008, Ramaswamy et al. 2010), although given the structural diversity of antimicrobials, degradation rates likely vary among compounds. A recent USDA study found that concentrations of extractable oxytetracycline in beef cattle manure mixed with straw and wood chips decreased by over 99% during 35 days of composting (Arikan et al. 2007). Additionally, populations of oxytetracycline-resistant bacteria were ten times lower in the manure after composting. This study suggests that adding straw and wood chips to manure, thereby increasing the temperature during composting, may allow for more rapid antimicrobial and pathogen reduction and/or adsorption. Arikan et al. (2009) documented declines of 99% and 98% in concentrations of extractable chlortetracycline and *epi*-chlortetracycline, respectively, in composted and sterile incubated manure mixtures. In another study, rates of antimicrobial decline in turkey litter extracts were measured during manure stockpiling, managed composting (i.e., routine mixing and managed moisture content), and in-vessel composting (i.e., controlled composting in a rotating steel drum) (Dolliver et al. 2008). In that study, chlortetracycline concentrations rapidly declined during all three treatments, with more than 99% removal within ten days. Concentrations of monensin and tylosin also decreased, but more gradually, with reductions ranging from 54% to 76% during the three treatments. In contrast, concentrations of sulfamethazine remained stable during all three treatments (Dolliver et al. 2008). In combination with recent research indicating that sulfonamides may be the most mobile antimicrobials (Chee-Sanford et al. 2009), the persistence of sulfamethazine (a type of sulfonamide) merits further study of its environmental occurrence and potential effects.

Composting is presumed to be an effective means of reducing hormone concentrations in manure via aerobic digestion (Zhao et al. 2008), though limited research has been conducted. One USDA study found that concentrations of 17β-estradiol and testosterone decreased by 84% and 90%, respectively, in chicken layer manure during composting (Hakk et al. 2005). In that study, testosterone concentrations declined at a faster rate than the 17β-estradiol concentrations. A more recent USDA study reported degradation of 17β-estradiol in poultry litter composted under heated conditions and at room temperature (Hakk et al. 2011). Limited research in this area is available, however, and further research on the degradation and adsorption of both natural and synthetic hormones in manure from various animal types would help elucidate the effectiveness of composting in removing hormones.

### 8.3.2.2. Anaerobic Digesters/Methane Capture

Anaerobic digesters, or biogas recovery systems, are oxygen-free environments in which bacteria break down manure, generating gases that may be captured for energy use. One of the primary gaseous byproducts of anaerobic digestion, methane, is combustible and may be used to generate electricity needs on the farm (e.g., to warm on-site buildings or heat water), sold to a local electric utility, or converted to compressed natural gas for fueling needs (USEPA 2011b). Liquid effluent from the digester may be spread on fields as fertilizer, since the digester does not reduce the nutrients in the manure. Digested solids may be used as livestock bedding material, or they may be sold for use as a soil amendment or for use in building materials such as particle board (USEPA 2011b).

There are a variety of types of anaerobic digesters; in 2010, the most commonly used types in the U.S. were mixed plug flow digesters (54%), complete mix digesters (42%), and covered lagoons (27%) (USEPA 2011c). A plug flow digester is a long, narrow, covered concrete tank and is used at dairy facilities that collect manure through scraping. A complete mix digester is an enclosed heated tank with a gas mixing system; this type of digester is optimal when manure is diluted with water. A covered lagoon digester is a lagoon with a flexible cover that minimizes atmospheric gas exchange and allows the recovered gas to be piped to a combustion device (USEPA 2011b).

The number of digesters in the U.S. has been steadily increasing since 2000 (USEPA 2011c). In 2010, there were 162 anaerobic digesters in the U.S., generating over 450 million kilowatt hours (kWh) of energy; this is equivalent to the amount of energy used to power 25,000 average American homes for a year. Additionally, the amount of methane emissions avoided due to use of digesters in 2010 was equivalent to reducing annual oil consumption by nearly 2.8 million barrels (USEPA 2011c). The majority of digesters are on dairy farms in the Midwest and Northeast, with 33 states having digesters in 2010 (USEPA 2011c).

The benefits of using anaerobic digesters include reductions in pathogens, reduced greenhouse gas emissions (methane and carbon dioxide), and minimization of odors (USDA 2011c). As reviewed by Sahlström (2003), while time and temperature (among other factors) influence pathogen inactivation, anaerobic digestion has been

---

**Anaerobic Digester Provides Farm a Source of Income and Reduces Environmental Impact:**

Brubaker Dairy Farms in Pennsylvania was named the 2011 Innovative Dairy Farmer of the Year by the International Dairy Foods Association for implementing an anaerobic digester powered by solar panels. The farm has over 1,400 cows and also produces 250,000 broilers annually. The digester kills fly larvae and weed seeds, reduces odors by 75% to 90%, and reduces the farm's methane and other greenhouse gas emissions.

All undigested fibers are reused as bedding for the cows or sold to other dairy farmers for bedding or gardening. The digester also generates enough energy in the form of electricity to power 150 to 200 homes per day. The majority of the energy is sold to a local utility, generating more income for the farm. Brubaker Dairy Farms has shown that these systems can work to minimize environmental impact and increase profit margin. (References: Brubaker 2009, IDFA 2011).

---

shown to be effective in reducing 90% of viable counts of microorganisms in hours (120-130°F) to days (86-100°F). Limited available research also suggests that anaerobic digesters may facilitate hormone and antimicrobial degradation. For example, concentrations of 17β-estradiol in dairy manure have been shown to decrease by 40% during anaerobic digestion (Zhao et al. 2008). A separate USDA experiment found that concentrations of oxytetracycline decreased by nearly 60% during 56 days in an anaerobic digester (Arikan et al. 2006). The study also reported that manure laden with 62 µg/g oxytetracycline and diluted 5-fold with water resulted in a 27% decrease in biogas production, indicating potential consequences of antibiotic use on the cost-effectiveness of anaerobic digestion. Levels of chlortetracycline in swine manure and monensin in cattle manure were also reduced by varying degrees after 21 days in anaerobic digesters set at different temperatures (Varel et al. 2012).

## 8.4. Financial and Technical Assistance Programs

Financial and technical assistance programs are available to help offset the costs of manure management. The table below highlights a few of the key federal programs managed by NRCS that provide financial assistance to producers. In addition to these resources, there are many state and local programs that provide loans and grants for reducing the environmental risks associated with manure.

**Table 8-1. Key USDA-NRCS programs that may provide financial assistance to producers.**

| Program Name | Description | Website |
|---|---|---|
| Agricultural Management Assistance (AMA) | Provides financial and technical assistance to agricultural producers to voluntarily address issues such as water management, water quality, and erosion control by incorporating conservation into their farming operations. | http://www.nrcs.usda.gov/wps/portal/nrcs/main/national/programs/financial/ama/ |
| Agricultural Water Enhancement Program (AWEP) | Voluntary conservation initiative that provides financial and technical assistance to agricultural producers to implement agricultural water enhancement activities on agricultural land to conserve surface and ground water and improve water quality. | http://www.nrcs.usda.gov/wps/portal/nrcs/detail/national/programs/financial/awep/?&cid=nrcs143_008334 |
| Conservation Innovation Grants (CIG) | Voluntary program intended to stimulate the development and adoption of innovative conservation approaches and technologies while leveraging Federal investment in environmental enhancement and protection, in conjunction with agricultural production. | http://www.nrcs.usda.gov/wps/portal/nrcs/detail/national/programs/financial/cig/?&cid=nrcs143_008205 |
| Environmental Quality Incentives Program (EQIP) | Voluntary program that provides financial and technical assistance to agricultural producers through contracts up to a maximum term of ten years in length. | http://www.nrcs.usda.gov/wps/portal/nrcs/main/national/programs/financial/eqip/ |

*Source: NRCS, 2012.* http://www.nrcs.usda.gov/wps/portal/nrcs/main/national/programs/financial

## 8.5. CAFO Regulations

The USGAO (2011b) noted that discharges from CAFOs share many of the traits of a diffuse, nonpoint source but are treated and regulated as a point source. The Clean Water Act specifically includes the term "concentrated animal feeding operation" in the definition of point source (Clean Water Act, Section 502(14)), and the NPDES program regulates discharges of pollutants from point sources. Under the NPDES permitting program, regulations governing CAFOs consist primarily of two different sets. The regulations at 40 CFR 122.23 set the framework for CAFO permitting by establishing criteria that define CAFOs and specifying whether and when a CAFO must have permit coverage. The second set of regulations, which are at 40 CFR Part 412, are the effluent limitations guidelines and standards for CAFOs, which establish discharge limits and certain management practice requirements that must be included in NPDES permits for CAFOs.

Any CAFO seeking NPDES permit coverage must submit a nutrient management plan as part of its permit application to be covered by an individual permit or a notice of intent to be covered by a general permit (40 CFR 122.23(h) and 122.42(e)(1)). A nutrient management plan is a manure and wastewater management tool that every permitted CAFO must use to properly manage discharges from the production or land application areas through the use of best management practices.

For more detailed information on CAFO regulations, refer to USEPA's CAFO rule history website: http://cfpub.epa.gov/npdes/afo/aforule.cfm. For further information on aquaculture NPDES regulations, visit: http://water.epa.gov/scitech/wastetech/guide/aquaculture/index.cfm.

## 8.6. Additional Technical Resources

A sampling of available on-line resources that are obtainable to help planners and producers related to manure management are listed in Appendix 3.

This page intentionally left blank.

# References

Aarestrup F.M., F. Bager, and J.S. Andersen. 2000. Association between the use of avilamycin for growth promotion and the occurrence of resistance among *Enterococcus faecium* from broilers: epidemiological study and changes over time. Microbial Drug Resistance. 6:71-75.

Abbaszadegan, M., M. LeChevallier, and C. Gerba. 2003. Occurrence of viruses in U.S. groundwaters. Journal of the American Water Works Association. 95:107-120.

Adams, C., Y. Wang, K. Loftin, and M. Meyer. 2002. Removal of antibiotics from surface and distilled water in conventional water treatment processes. Journal of Environmental Engineering. 128(3):253-260.

Adams, C. 2008. Treatment of antibiotics in swine wastewater. p. 331-348. *In* D.S. Aga (ed.) Fate and transport of pharmaceuticals in the environment and water treatment systems. 1st ed. CRC Press, Boca Raton, FL.

Ainsworth, L., R.H. Common, and R.L. Carter. 1962. A chromatographic study of some conversion products of estrone-16-C-14 in the urine and feces of the laying hen. Canadian Journal of Biochemistry and Physiology. 40:123-135.

American Water Works Association (AWWA). 1999. Waterborne pathogens. 1st ed. AWWA Manual M48, Glacier Publishing Services, Denver, CO.

Amirkolaie, A.K. 2011. Reduction in the environmental impact of waste discharged by fish farms through feed and feeding. Reviews in Aquaculture. 3:19-26.

Aneja, V.P., B. Bunton, J.T. Walker, and B.P. Malik. 2001. Measurement and analysis of atmospheric ammonia emissions from anaerobic lagoons. Atmospheric Environment. 35:1949-1958.

Animal Health Institute (AHI). 2000. Survey indicates most antibiotics used in animals are used for treating and preventing disease. Animal Health Institute, Washington, DC.

Ankley, G.T., K.M. Jensen, E.A. Makynen, M.D. Kahl, J.J. Korte, M.W. Hornung, T.R. Henry, J.S. Denny, R.L. Leino, V.S. Wilson, M.C. Cardon, P.C. Hartig, and L.E. Gray. 2003. Effects of the androgenic growth promoter 17-β-trenbolone on fecundity and reproductive endocrinology of the fathead minnow. Environmental Toxicology and Chemistry. 22(6):1350-1360.

Apley, M. 2004. Importance of macrolides in veterinary medicine. Veterinary Diagnostic and Production animal medicine, Iowa State University. Presented at the Veterinary Medicine Advisory Committee meeting. October 13, 2004. Rockville, MD. www.fda.gov/downloads/AdvisoryCommittees/.../UCM129178.ppt. (Verified, December 2011).

Arikan, O.A., L.J. Sikora, W. Mulbry, S.U. Khan, C. Rice, and G.D. Foster. 2006. The fate and effect of oxytetracycline during the anaerobic digestion of manure from therapeutically treated calves. Process Biochemistry. 41:1637-1643.

Arikan, O.A., L.J. Sikora, W. Mulbry, S.U. Khan, and G.D. Foster. 2007. Composting rapidly reduces levels of extractable oxytetracycline in manure from therapeutically treated beef calves. Bioresource Technology. 98:169-176.

Arikan, O.A., W. Mulbry, and C. Rice. 2009. Management of antibiotic residues from agricultural sources: use of composting to reduce chlortetracycline residues in beef manure from treated animals. Journal of Hazardous Materials. 164:483-489.

Armstrong, J. L., D. S. Shigeno, J. J. Calomiris, and R. J. Seidler. 1981. Antibiotic-resistant bacteria in drinking water. Applied Environmental Microbiology. 42:277–283.

Armstrong, J. L., J. J. Calomiris, and R. J. Seidler. 1982. Selection of antibiotic-resistant standard plate count bacteria during water treatment. Applied Environmental Microbiology. 44:308–316.

Armstrong, S.D., D.R. Smith, B.C. Joern, P.R. Owens, A.B. Leytem, C. Huang, and L. Adeola. 2010. Transport and fate of phosphorus during and after manure spill simulations. Journal of Environmental Quality. 39:345-352.

Arnon, S., O. Dahan, S. Elhanany, K. Cohen, I. Pankratov, A. Gross, Z. Ronen, S. Baram, and L.S. Shore. 2008. Transport of testosterone and estrogen from dairy-farm waste lagoons to groundwater. Environmental Science & Technology. 42(15):5521-5526.

Ash, R., B. Mauck, and M. Morgan. 2002. Antibiotic resistance of gram-negative bacteria in rivers, United States. Emerging Infectious Diseases. 8:713-716.

Atherton, F.C., P.S., Newman, and D.P. Casemore. 1995. An outbreak of waterborne cryptosporidiosis associated with a public water supply in the UK. Epidemiology and Infection. 115:123-131.

Atwill, E. 1995. Microbial pathogens excreted by livestock and potentially transmitted to humans through water. Veterinary Medicine Teaching and Research Center, School of Veterinary Medicine, University of California, Davis.

Banks, M.K., W. Yu, and R.S. Govindaraju. 2003. Bacterial adsorption and transport in saturated soil columns. Journal of Environmental Science and Health, Part A: Toxic/Hazardous Substances and Environmental Engineering. 38(12):2749-2758.

Bartelt-Hunt, S., D.D. Snow, W.L. Kranz, T.L. Mader, C.A. Shapiro, S.J. van Donk, D.P. Shelton, D.D. Tarkalson, and T.C. Zhang. 2012. Effect of growth promotants on the occurrence of endogenous and synthetic steroid hormones on feedlot soils and in runoff from beef cattle feeding operations. Environmental Science & Technology. 46(3):1352-1360.

Batt, A.L., D.D. Snow, and D.S. Aga. 2006. Occurrence of sulfonamide antimicrobials in private water wells in Washington County, Idaho, USA. Chemosphere. 64(11):1963-1971.

Battigelli, D. A., M. D. Sobsey, and D. C. Lobe. 1993. The inactivation of hepatitis A virus and other model viruses by UV irradiation. Water Science and Technology. 27:339-342.

Baxter-Potter, W., and M. Gilliland. 1988. Bacterial pollution in runoff from agricultural lands. Journal of Environmental Quality. 17:27-34.

Becker, G.S. 2010. Antibiotic use in agriculture: background and legislation. Congressional Research Service Report for Congress. 7-5700, R40739.

Benbrook, C.M. 2001. Quantity of antimicrobials used in food animals in the United States. American Society for Microbiology 101st Annual Meeting. May 20-24, 2001. Orlando, FL.

Benbrook, C.M. 2002. Antimicrobial drug use in U.S. aquaculture. The Northwest Science and Environmental Policy Center. Sandpoint, Idaho.

Benotti, M.J., R.A. Trenholm, B.J. Vanderford, J.C. Holady, B.D. Stanford, and S.A. Snyder. 2009. Pharmaceuticals and endocrine disrupting compounds in U.S. drinking water. Environmental Science & Technology. 43:597-603.

Bevacqua, C.E., C.P. Rice, A. Torrents, and M. Ramirez. 2011. Steroid hormones in biosolids and poultry litter: a comparison of potential environmental inputs. Science of the Total Environment. 409(11):2120-2126.

Bicudo, J.R., and S. Goyal. 2003. Pathogens and manure management systems: a review. Environmental Technology. 24(1):115-130.

Blazer, V.S., L.R. Iwanowicz, D.D. Iwanowicz, D.R. Smith, J.A. Young, J.D. Hedrick, S.W. Foster, and S.J. Reeser. 2007. Intersex (testicular oocytes) in smallmouth bass from the Potomac River and selected nearby drainages. Journal of Aquatic Animal Health. 19:242-253.

Bopp, D., B.D. Sauders, A.T. Waring, J. Ackelsberg, N. Dumas, E Braun-Howland, D. Dziewulsky, B.J. Wallace, M. Kelly, T. Halse, K. Aruda Musser, P.F. Smith, D.L. Morse, and R.J. Limberger. 2003. Detection, isolation, and molecular subtyping of *Escherichia coli* O157:H7 and *Campylobacter* jejuni associated with a large waterborne outbreak. Journal of Clinical Microbiology. 41(1):174-180.

Borchardt, M.A., K.R. Bradbury, M.B. Gotkowitz, J.A. Cherry, and B.L. Parker. 2007. Human enteric viruses in groundwater from a confined bedrock aquifer. Environmental Science & Technology. 41(18):6606-6612.

Bosch, A., R. Pinto, and F. Abad. 2006. Survival and transport of enteric viruses in the environment. p. 151-187. *In* S.M. Goyal (ed.) Viruses in Foods. Food Microbiology and Food Safety.

Bouwman, A.F., D.S. Lee, W.A.H. Asman, F.J. Dentener, K.W. van der Hoek, and J.G.J. Olivier. 1997. A global high-resolution emission inventory for ammonia. Global Biogeochemical Cycles. 11(4):561-587.

Bowman, J. 2009. Manure pathogens: manure management, regulations, and water quality protection. p. 562. Water Environmental Federation, McGraw-Hill, New York.

Boxall, A., D. Kolpin, B. Holling-Sorensen, and J. Tolls. 2003. Are veterinary medicines causing environmental risks? Environmental Science & Technology. 37(15):286a-294a.

Boxall, A. 2008. Fate and transport of veterinary medicines in the soil environment. p 123-137. *In* D.S. Aga (ed.) Fate and transport of pharmaceuticals in the environment and water treatment systems. 1st ed. CRC Press, Boca Raton, FL.

Bradford, S. and J. Schijven. 2002. Release of *Cryptosporidium* and *Giardia* from dairy calf manure: impact of solution salinity. Environmental Science & Technology. 36(18):3916-3923.

Brewer, R.A. and M.J. Corbel. 1983. Characterization of Yersinia enterocolitica strains isolated from cattle, sheep and pigs in the United Kingdom.. J Hyg (Lond). 90(3):425-33.

Brookes, J., J. Antenucci, M. Hipsey, M. Burch, M., N. Ashbolt, and C. Ferguson. 2004. Fate and transport of pathogens in lakes and reservoirs. Environment International. 30:741-759.

Brubaker, L. 2009. Producer perspectives as they relate to dairy farms & global warming. Senate Testimony to Chairman Harkin, Senator Casey and Agriculture Committee Members. September 16, 2009. http://ag.senate.gov/download/brubaker-testimony. (Verified December, 2011).

Burgos, J.M., B.A. Ellington, and M.F. Varela. 2005. Presence of multidrug resistant enteric bacteria in dairy farm topsoil. Journal of Dairy Science. 88(4):1391-1398.

Camargo, J.A., and A. Alonso. 2006. Ecological and toxicological effects of inorganic nitrogen pollution in aquatic ecosystems: a global assessment. Environment International. 32:831-849.

Campagnolo, E.R., K.R. Johnson, A. Karpati, C.S. Rubin, D.W., Kolpin, M.T. Meyer, J.E. Esteban, R.W. Currier, K. Smith, K. M. Thu, and M. McGeehin. 2002. Antimicrobial residues in animal waste and water resources proximal to large-scale swine and poultry feeding operations. The Science of the Total Environment. 299:89-95.

Caron, E., A. Farenhorst, F. Zvomuya, J. Gaultier, N. Rank, T. Goddard, and C. Sheedy. 2010. Sorption of four estrogens by surface soils from 41 cultivated fields in Alberta, Canada. Geoderma. 155:19-30.

Carmosini, N., and L.S. Lee. 2008. Sorption and Degradation of selected pharmaceuticals in soil and manure. p 139-165. *In* D.S. Aga (ed.) Fate and transport of pharmaceuticals in the environment and water treatment systems. 1st ed. CRC Press, Boca Raton, FL.

Casey, F.X.M., H. Hakk, J. Šimunek, and G.L. Larsen. 2004. Fate and transport of testosterone in agricultural soils. Environmental Science & Technology. 38:790-798.

Centers for Disease Control and Prevention (CDC). 1999. Outbreak of *Escherichia coli* O157:H7 and *Campylobacter* among attendees of the Washington County Fair- New York. Morbidity and Mortality Weekly Report, 48: 803–805, September 17, 1999.

CDC 2011a. 2011. CDC Estimates of Foodborne Illness in the United States: Clostridium perfringens. http://www.cdc.gov/foodborneburden/clostridium-perfringens.html (Verified January 2013).

CDC. 2011b. Summary of notifiable diseases, United States. Morbidity and Mortality Weekly Report, 58(53), May 13, 2011.

CDC. 2013. National Center for Emerging and Zoonotic Infectious Diseases: our work, our stories, 2011-2012. Atlanta, GA.

Characklis, G.W., M.J. Dilts, O.D. Simmons III, C.A. Likirdopulos, L.A.H. Krometis, and M.D. Sobsey. 2005. Microbial partitioning to settleable particles in stormwater. Water Research. 39:1773-1782.

Chee-Sanford, J.C., R.I. Aminov, I.J. Krapac, N. Garrigues-Jeanjean, and R.I. Mackie. 2001. Occurrence and diversity of tetracycline resistance genes in lagoons and groundwater underlying two swine production facilities. Applied Environmental Microbiology. 67(4):1494-1502.

Chee-Sanford, J.C., R.I. Mackie, S. Koike, I.G. Krapac, Y. Lin, A.C. Yannarell, S. Maxwell, and R.I. Aminov. 2009. Fate and transport of antibiotic residues and antibiotic resistance genes following land application of manure waste. Journal of Environmental Quality. 38(3):1086-1108.

Chemaly M, M T Toquin, Y Le Nôtre, P Fravalo. 2008. Prevalence of Listeria monocytogenes in poultry production in France. Journal of Food Protection. 71(10): 1996-2000.

Chesapeake Bay Commission. 2012. Manure to energy: sustainable solutions for the Chesapeake Bay region. January 2012. http://www.chesbay.us/Publications/manure-to-energy%20report.pdf (Verified June 2013).

Cho, K., Y. Pachepsky, J. Ha Kim, A. Guber, D. Shelton, and R. Rowland. 2010. Release of *Escherichia coli* from the bottom sediment in a first-order creek: experiment and reach-specific modeling. Journal of Hydrology. 391:322–332.

Choi, H.S., E. Kiesenhofer, H. Gantner, J. Hois, and E. Bamberg. 1987. Pregnancy diagnosis in sows by estimation of oestrogens in blood, urine or faeces. Animal Reproduction Science. 15:209-216.

Cieslak, P.R., T.J. Barrett, P.M. Griffin, K.F. Gensheimer, G. Beckett, J. Buffington, and M.G. Smith. 1993. *Escherichia coli* 0157:H7 Infection from a Manured Garden. The Lancet. 342(8867):342-367.

Ciparis, S., L.R. Iwanowicz, and J.R. Voshell. 2012. Effects of watershed densities of animal feeding operations on nutrient concentrations and estrogenic activity in agricultural streams. Science of the Total Environment. 414:268-276.

Cizek, A.R., G.W. Characklis, L.A.H. Krometis, J.A. Hayes, O.D. Simmons III, S. DiLonardo, K.A. Alderisio, and M.D. Sobsey. 2008. Comparing the partitioning behavior of *Giardia* and *Cryptosporidium* with that of indicator organisms in stormwater runoff. Water Research. 42:4421-4438.

Clark, C.G., L. Price, R. Ahmed, D.L. Woodward, P.L. Melito, F.G. Rodgers, F. Jamieson, B. Ciebin, A. Li, and A. Ellis. 2003. Characterization of waterborne outbreak-associated *Campylobacter jejuni*, Walkerton, Ontario. Emerging Infectious Diseases. 9(10):1232-1241.

Cole, D.W., R. Cole, S.J. Gaydos, J. Gray, G. Hyland, M.L. Jacques, N. Powell-Dunford, C. Sawhney, and W.W. Au. 2009. Aquaculture: environmental, toxicological, and health issues. International Journal of Hygiene and Environmental Health. 212:369-377. Cook, N., J. Bridger, K. Kendall, M.I Gomara, L. El-Attar, and J. Gray. 2004. The zoonotic potential of rotavirus. Journal of Infection. 48(4):289-302.

Cotruvo, J.A., A. Dufour, G. Rees, J. Bartram, R. Carr, D.O. Cliver, G.F. Craun, R. Fayer, and V.P.J. Gannon. 2004. Waterborne Zoonoses: Identification, Causes and Control. World Health Organization, IWA Publishing, London, UK.

Cox, N.A., N.J. Stern, M.T. Musgrove, J.S. Bailey, S.E. Craven, P.F. Cray, R.J. Buhr, and K.L. Hiett. 2002. Prevalence and level of *campylobacter* in commercial broiler breeders (parents) and broilers. The Journal of Applied Poultry Research. 11:187-190.

Crabill, C., R. Donald, J. Snelling, R. Foust, R., and G. Southam. 1999. The impact of sediment fecal coliform reservoirs on seasonal water quality in Oak Creek, Arizona. Water Research. 33(9):2163-2171.

Cransberg, K., J.H. van den Kerkhof, J.R. Banffer, C. Stijnen, K. Wernars, N.C. van de Kar, J. Nauta, and E.D. Wolff. 1996. Four cases of hemolytic uremic syndrome- source contaminated swimming water? Clinical Nephrology. 46:45-49.

Craun, M.F., G.F. Craun, R.L. Calderon, and M.J. Beach. 2006. Waterborne outbreaks reported in the United States. Journal of Water and Health. 4(2):19-30.

Curriero, F.C., J.A. Patz, J.B. Rose, and S. Lele. 2001. The association between extreme precipitation and waterborne disease outbreaks in the United States, 1948-1994. American Journal of Public Health. 91(8):1194-1199.

Dargatz, D. 1996. Shedding of *Escherichia coli* O157:H7 by feedlot cattle. p. 1.7.1-1.7.3. *In* APHIS report of accomplishments in Animal Production Food Safety FY 1995/1996. USDA, Animal and Plant Health Inspection Service.

Daughton, C.G., and T.A. Ternes. 1999. Pharmaceuticals and personal care products in the environment: agents of subtle change? Environmental Health Perspectives. 107:907-938.

Davies-Colley, R., J. Nagels, A. Donnison, and R. Muirhead. 2004. Flood flushing of bugs in agricultural streams. Water and Atmosphere 12(2):18-20.

DeVore, B. 2002. Manure matters. Minnesota Conservation Volunteer Magazine. Minnesota Department of Natural Resources, July-August. http://www.dnr.state.mn.us/volunteer/julaug02/feedlots.htmL. (Verified December, 2011).

Dewey, C.E., B.D. Cox, B.E. Straw, E.J. Budh, and H.S. Hurd. 1997. Association between off-label feed additives and farm size, veterinary consultant use, and animal age. Preventative Veterinary Medicine. 31:133-146.

Dodd, M.C., A.D. Shah, U. Von Gunten, and C.H. Huang. 2005. Interactions of fluoroquinolone antibacterial agents with aqueous chlorine: reaction kinetics, mechanisms, and transformation pathways. Environmental Science & Technology. 39(18):7065-7076.

Dolliver, H., S. Gupta, and S. Noll. 2008. Antibiotic degradation during manure composting. Journal of Environmental Quality. 37:1245-1253.

Donham K., S. Reynolds, P. Whitten, J. Merchant, L. Burmeister, and W. Popendorf. 1995. Respiratory dysfunction in swine production facility workers: dose-response relationships of environmental exposures and pulmonary function. American Journal of Industrial Medicine. 27:405-418.

Donham, R.J., S. Wing, D. Osterberg, J.L. Flora, C. Hodne, K.M. Thu, and P.S. Thorne. 2007. Community health and socioeconomic issues surrounding concentrated animal feeding operations. Environmental Health Perspectives. 115(2):317-320.

Dubrovsky, N.M., K.R. Burow, G.M. Clark, J.M. Gronberg, P.A. Hamilton K.J. Hitt, D.K. Mueller, M.D. Munn, B.T. Nolan, L.J. Puckett, M.G. Rupert, T.M. Short, N.E. Spahr, L.A. Sprague, and W.G. Wilber. 2010. The quality of our nation's waters- nutrients in the nation's streams and groundwater, 1992–2004. U.S. Geological Survey Circular 1350, USGS, Reston, VA.

Durhan, E.J., C.S. Lambright, E.A. Makynen, J. Lazorchak, P.C. Hartig, V.S. Wilson, L.E. Gray, and G.T. Ankley. 2006. Identification of metabolites of trenbolone acetate in androgenic runoff from a beef feedlot. Environmental Health Perspectives. 114:65-68.

Edzwald, J.K. (ed.) 2010. Water quality and treatment: a handbook on drinking water. 6th ed. American Water Works Association (AWWA).

Emborg, H.D., J.S. Andersen, A.M. Seyfarth, S.R. Andersen, J. Boel, and H.C. Wegener. 2003. Relations between the occurrence of resistance to antimicrobial growth promoters among *Enterococcus faecium* isolated from broilers and broiler meat. International Journal of Food Microbiology. 84:273-284.

Europa. 2005. Ban on antibiotics as growth promoters in animal feed enters into effect. Europa Press Releases RAPID, 22 December 2005. http://europa.eu/rapid/pressReleasesAction.do?reference=IP/05/1687&format=HTML&aged=0&language=EN&guiLanguage=en. (Verified December, 2011).

Europa. 2011. Action plan against antimicrobial resistance: commission unveils 12 concrete actions for the next five years. http://europa.eu/rapid/pressReleasesAction.do?reference=IP/11/1359. (Verified February, 2012).

Feinman, S.E., and J.C. Matheson, III. 1978. Draft environmental impact statement: subtherapeutic antibacterial agents in animal feeds. USFDA, Bureau of Veterinary Medicine, Rockville, MD.

Feldman, K.A., J.C. Mohle-Boetani, J. Ward, K. Furst, S.L. Abbott, D.V. Ferrero, A. Olsen, and S.B. Werner. 2002. A cluster of *Escherichia coli* O157: nonmotile infections associated with recreational exposure to lake water. Public Health Reports. 117(4):380–385.

Fernández-Barredo, S., C. Galiana, A. García, S. Vega, M. T. Gómez, and M.T. Pérez-Gracia. 2006. Detection of hepatitis E virus shedding in feces of pigs at different stages of production using reverse transcription-polymerase chain reaction. Journal of Veterinary Diagnostic Investigation. 18: 462-465.

Fong, T.T., and E.K. Lipp. 2005. Enteric viruses of humans and animals in aquatic environments: health risks, detection, and potential water quality assessment tools. Microbiology and Molecular Biology Reviews. 69(2):357-371.

Fujioka, R.S., and B.S. Yoneyama. 2002. Sunlight inactivation of human enteric viruses and fecal bacteria. Water Science and Technology. 46:291-295.

Fukushima, H., T. Hashizume, Y. Morita, J. Tanaka, K. Azuma, Y. Mizumoto, M. Kaneno, M. Matsuura, K. Konma, and T. Kitani. 1999. Clinical experiences in Sakai City Hospital during the massive outbreak of enterohemorrhagic *Escherichia coli* O157 infections in Sakai City, 1996. Pediatrics International. 41:213-217.

Garber, L., H.S. Hurd, T. Keefe, and J.L. Schlater. 1994. Potential risk factors for *Cryptosporidium* infection in dairy calves. Journal of the American Veterinary Medical Association. 205(1):86-91.

Geldreich, E., K. Fox, J. Goodrich, E. Rice, R. Clark, and D. Swerdlow. 1992. Searching for a water supply connection in the Cabool, Missouri disease outbreak of *Escherichia coli* O157:H7. Water Research Journal. 26(8):1127-1137.

Gelting, R. 2006. Investigation of an *Escherichia coli* O157:H7 outbreak associated with Dole pre-packaged spinach. Attachment 11: CDC Addendum Report, "Irrigation Water Issues Potentially Related to 2006 E. coli O157:H7 in Spinach Outbreak." http://www.cdc.gov/nceh/ehs/Docs/Investigation_of_an_E_Coli_Outbreak_Associated_with_Dole_Pre-Packaged_Spinach.pdf. (Verified March, 2012).

Gerba, C.P., and J.E. Smith. 2005. Sources of pathogenic microorganisms and their fate during land application of wastes. Journal of Environmental Quality. 34:42-48.

Gibbs, S.G., C.F. Green, P.M. Tarwater, and P.V. Scarpino. 2004. Airborne antibiotic resistant and nonresistant bacteria and fungi recovered from two swine herd confined animal feeding operations. Journal of Occupational and Environmental Hygiene. 1:699-706.

Gibson, R. 2011. When nutrients run amuck: the Grand Lake dilemma. Ohio Environmental Protection Agency. Presented at the USEPA Nutrient TMDL Workshop. February 15-17, 2011. New Orleans, LA. http://www.epa.gov/Region5/agriculture/pdfs/nutrientworkshop/07gibson.pdf. (Verified December, 2011).

Gilbert, N. 2012. Rules tighten on use of antibiotics on farms. Nature. 481:125.

Giri, R.R., H. Ozaki, Y. Takayanagi, S. Taniguchi, and R. Takanami. 2011. Efficacy of ultraviolet radiation and hydrogen peroxide oxidation to eliminate large number of pharmaceutical compounds in mixed solution. International Journal of Environmental Science & Technology. 8(1):19-30.

Gollehon, N., M. Caswell, M. Ribaudo, R. Kellogg, C. Lander, and D. Letson. 2001. Confined animal production and manure nutrients. Agriculture Information Bulletin No. 771. USDA, NRCS, ERS, Washington, DC.

Goyal, S.M., and C.P. Gerba. 1979. Comparative adsorption of human enteroviruses, simian rotavirus, and selected bacteriophages to soils. Applied and Environmental Microbiology. 38:241-247.

Graham, J.P., and K.E. Nachman. 2010. Managing waste from confined feeding operations in the U.S. Journal of Water and Health. 8(4):646-670.

Gray, T.W. 2006. Dairy dilemma: ban on rBGH use by Tillamook sparks conflict. Rural Cooperatives, November 2006. USDA, Rural Development, Cooperative Programs. http://www.rurdev.usda.gov/rbs/pub/nov06/dairy.htm. (Verified February, 2012).

Grieg, J.D., and E.C.D. Todd. 2010. Infective doses and pathogen carriage. Public Health Agency of Canada. Presented at the 2010 Food Safety Education Conference. March 25, 2010. Atlanta, GA.

Gu, C., and K.G. Karthikeyan. 2005. Sorption of the antimicrobial ciprofloxacin to aluminum and iron hydrous oxides. Environmental Science & Technology. 39:9166-9173.

Guan, T.Y., and R.A. Holley. 2003. Pathogen survival in swine manure environments and transmission of human enteric illness- a review. Journal of Environmental Quality. 32:383-392.

Gullick, R.W., R.A. Brown, and D.A. Cornwell. 2007. Source water protection for concentrated animal feeding operations: a guide for drinking water utilities. Awwa Research Foundation and the U.S. Environmental Protection Agency. IWA Publishing.

Hakk, H., P. Millner, and G. Larsen. 2005. Decrease in water-soluble 17β-estradiol and testosterone in composted poultry manure with time. Journal of Environmental Quality. 34:943-950.

Hakk, H., and L. Sikora. 2011. Dissipation of 17β-estradiol in composted poultry litter. Journal of Environmental Quality. 40:1560-1566.

Hancock, D.D., D.H. Rice, L.A. Thomas, and D.A. Dargatz. 1997. Epidemiology of *Escherichia coli* O157 in feedlot cattle. Journal of Food Protection. 60:462-465.

Hanselman, T.A., D.A. Graetz, and A.C. Wilkie. 2003. Manure-borne estrogens as potential environmental contaminants: a review. Environmental Science & Technology. 37:5471-5478.

Harden, S.L. 2009. Reconnaissance of organic wastewater compounds at a concentrated swine feeding operation in the North Carolina coastal plain, 2008. USGS and the North Carolina Department of Environment and Natural Resources, Division of Water Quality, Aquifer Protection Section. USGS Open File Report 2009-1128.

Harrington, G.W., H. Chen, A.J. Harris, I. Xagoraraki, D.A. Battigelli, and J.H. Standridge. 2001. Removal of Emerging Waterborne Pathogens. American Water Works Association Research Foundation, Denver.

Hayes, J.R., L.L. English, L.E. Carr, D.D. Wagner, and S.W. Joseph. 2004. Multiple-antibiotic resistance of enterococcus spp. isolated from commercial poultry production environments. Applied and Environmental Microbiology. 70(10):6005-6011.

Health Canada, 2009. Listeriosis Investigative Review. http://www.listeriosis-listeriose.investigation-enquete.gc.ca/index_e.php?s1=rpt&page=chap02 (Verified January, 2013).

Hendricks, D.W., F.J. Post, and D.R. Khairnar. 1979. Adsorption of bacteria onto soils. Water, Air, & Soil Pollution. 12:219-232.

Herrman, T. and G.L. Stokka. 2001. Medicated feed additives for beef cattle and calves. MF-2043- Feed Manufacturing. Kansas State University Agricultural Experimental Station and Cooperative Service.

Herrman, T., and P. Sundberg. 2001. Medicated feed additives for swine. MF-2042- Feed Manufacturing. Kansas State University, Agricultural Experimental Station and Cooperative Service.

Hilborn, E.D., J.H. Mermin, P.A. Mshar, J.L. Hadler, A. Voetsch, C. Wojtkunski, M. Swartz, R. Mshar, M.A. Lambert-Fair, J.A. Farrar, M.K. Glynn, and L. Slutsker. 1999. A multistate outbreak of *Escherichia coli* O157:H7 infections associated with consumption of mesclun lettuce. Archives of Internal Medicine. 159:1758-1764.

Hlavsa, M.C., V.A. Roberts, A.R. Anderson, V.R. Hill, A.M. Kahler, M. Orr, L.E. Garrison, L.A. Hicks, A. Newton, E.D. Hilborn, T.J. Wade, M.J. Beach, and J.S. Yoder. 2011. Surveillance for waterborne disease outbreaks and other health events associated with recreational water – United States, 2007-2008. Centers for Disease Control and Prevention, Morbidity and Mortality Weekly Report. 60(ss12):1-32. http://www.cdc.gov/mmwr/preview/mmwrhtml/ss6012a1.htm. (Verified March, 2012).

Ho AJ, R. Ivanek, Y T Gröhn, K K Nightingale, and Wiedmann M. 2007. Listeria monocytogenes fecal shedding in dairy cattle shows high levels of day-to-day variation and includes outbreaks and sporadic cases of shedding of specific L. monocytogenes subtypes. Prev Vet Med. 2007 Aug 16;80(4):287-305.

Hoffmann, B., T. Goes de Pinho, and G. Schuler. 1997. Determination of free and conjugated oestrogens in peripheral blood plasma, feces and urine of cattle throughout pregnancy. Experimental and Clinical Endocrinology & Diabetes. 105:296-303.Houf K, De Zutter L, Verbeke B, Van Hoof J, Vandamme P. Molecular characterization of Arcobacter isolates collected in a poultry slaughterhouse. J Food Prot. 2003;66:364–9.

Hoxie, N.J., J.P.Davis, J.M. Vergeron, R.D. Nashold, and K.A. Blair. 1997. Cryptosporidiosis associated mortality following a massive waterborne outbreak in Milwaukee, Wisconsin. American Journal of Public Health. 87:2032-2035.

Hrudey, S.E., P. Payment, P.M. Huck, R.W. Gillham, and E.J. Hrudey. 2003. A fatal waterborne disease epidemic in Walkerton, Ontario: comparison with other waterborne outbreaks in the developed world. Water Science and Technology. 47(3):7-14.

Hunter, P.R., Y Andersson, C.H. Von Bondsorff, R.M. Chalmers, E. Cifuentes, D. Deere, T. Endo, M. Kadar, T. Krogh, L. Newport, A. Prescott, and W. Robertson. 2003. Surveillance and Investigation of Contamination Incidents and Waterborne Outbreaks. p. 205-236. *In* A. Dufour, M. Snozzi, W. Koster, J. Bartam, E. Ronchi, and L. Fewtrell (ed.) Assessing Microbial Safety of Drinking Water: Improving Approaches and Methods. World Health Organization, IWA Publishing, London, UK.

Hurst, C.J., C.P. Gerba, and I. Cech. 1980. Effects of environmental variables and soil characteristics on virus survival in soil. Applied and Environmental Microbiology. 40(6):1067-1079.

Husu, J R. 1990. Epidemiological Studies on the Occurrence of Listeria monocytogenes in the Feces of Dairy Cattle. Journal of Veterinary Medicine, Series B. 37(1-10): 276–282.

Hutchison, M., L. Walters, S. Avery, F. Munro, and A. Moore. 2004. Analyses of livestock production, waste storage, and pathogen levels and prevalences in farm manures. Applied and Environmental Microbiology. 71(3):207-214.

Ihekweazu, C., M. Barlow, S. Roberts, H. Christensen, B. Guttridge, D. Lewis, and S. Paynter. 2006. Outbreak report: outbreak of E. coli O157 infection in the south west of the UK: risks from stream crossing seaside beaches. Eurosurveillance. 11(4):613.

International Dairy Foods Association (IDFA). 2011. Brubaker Farms of Mount Joy, Pennsylvania, named 2011 innovative dairy farmer of the year. January 24, 2011. http://www.idfa.org/news--views/headline-news/details/5573/. (Verified December, 2011).

Ivie, G.W., R.J. Christopher, C.E. Munger, and C.E. Copperpock. 1986. Fate and residues of [4-14C] estradiol-17 beta after intramuscular injection into Holstein steer calves. Journal of Animal Science. 62:681-690.

Iwanowicz, L.R., and V.S. Blazer. 2011. An overview of estrogen-associated endocrine in fishes: evidence of effects on reproductive and immune physiology. p. 266-275. *In* R.C. Cipriano, A. Bruckner, and I. S. Shchelkunov (ed.) Aquatic animal health: a continuing dialogue between Russia and the United States. Proceedings of the Third Bilateral Conference Between the United States and Russia: Aquatic Animal Health 2009. July 12-20, 2009. Shepherdstown, WV. Michigan State University, East Lansing, Michigan.

Jackson, C.R., P.J. Fedorka-Cray, J.B. Barrett, S.R. Ladeley. 2004. Effects of tylosin use on erythromycin resistance in enterococci isolated from swine. Applied and Environmental Microbiology. 70(7):4205-4210.

Jacobs, J.J. 2012. FDA: judge orders agency to review animal antibiotic risks. Environment and Energy Reporter. Tuesday, June 5, 2012.

James, R., M.L. Eastridge, L.C. Brown, K.H. Elder, S.S. Foster, J.J. Hoorman, M.J. Joyce, H.M. Keener, K. Mancl, M.J. Monnin, J.N. Rausch, J.M. Smith, O. Tuovinen, M.E. Watson, M.H. Wicks, N. Widman, and L. Zhao. 2006. Ohio livestock manure management guide. The Ohio State University Extension. Bulletin 604.

Jamieson, R.C., R.J. Gordon, K.E. Sharples, G.W. Stratton, and A. Madani, A. 2002. Movement and persistence of fecal bacteria in agricultural soils and subsurface drainage water: A review. Canadian Biosystems Engineering/Le génie des biosystèmes au Canada. 44:1.1-1.9.

Jawson, M.D., L.F. Elliott, K.E. Saxton, and D.H. Fortier. 1982. The effect of cattle grazing on indicator bacteria in runoff from a Pacific Northwest watershed. Journal of Environmental Quality. 11:621-627.

Jay MT, Cooley M, Carychao D, Wiscomb GW, Sweitzer RA, Crawford-Miksza L, et al. 2007. Escherichia coli O157:H7 in feral swine near spinach fields and cattle, central California coast. Emerg Infect Dis 2007 Dec. Available from http://wwwnc.cdc.gov/eid/article/13/12/07-0763.htm

Jensen, K.M., E.A. Makynen, M.D. Kahl, and G.T. Ankley. 2006. Effects of the feedlot contaminant 17$\alpha$-trenbolone on reproductive endocrinology of the fathead minnow. Environmental Science & Technology. 40:3112-3117.

Jordan, T.E., and D.E. Weller. 1996. Human contributions to terrestrial nitrogen flux: assessing the sources and fates of anthropogenic fixed nitrogen. Bioscience. 46(9):655-664.

Jürgens, M.D., K.I.E. Holthaus, A.C. Johnson, and J.J.L. Smith. 2002. The potential for estradiol and ethynylestradiol degradation in English rivers. Environmental Toxicology and Chemistry. 21:480-488.

Kapuscinski, R.B., and R. Mitchell. 1983. Sunlight-induced mortality of viruses and *Escherichia coli* in coastal seawater. Environmental Science & Technology. 17(1):1-6.

Kasorndorkbua, C., T. Opriessnig, F.F. Huang, D.K. Guenette, P.J. Thomas, X.J. Meng, and P.G. Halbur. 2005. Infectious swine hepatitis E virus is present in pig manure storage facilities on United States farming, but evidence of water contamination is lacking. Applied and Environmental Microbiology. 71(12):7831-7837.

Kellogg, R.L., C.H. Lander, D.C. Moffitt, and N. Gollehon. 2000. Manure nutrients relative to the capacity of cropland and pastureland to assimilate nutrients: spatial and temporal trends for the United States. Publication No. nps00-0579. USDA, NRSC, ERS, Washington, DC.

Kemp, R., A.J.H. Leatherbarrow, N.J. Williams, C.A. Hart, H.E. Clough, J. Turner, E.J. Wright, and N.P. French. 2005. Prevalence and genetic diversity of *Campylobacter* spp. in environmental water samples from a 100-square-kilometer predominantly dairy farming area. Applied and Environmental Microbiology. 71(4):1876-1882.

Khan S.J., D.J. Roser, C.M. Davies, G.M. Peters, R.M. Stuetz, R. Tucker, and N.J. Ashbolt. 2008. Chemical contaminants in feedlot wastes: concentrations, effects and attenuation. Environment International 34:839-859.

Khan, B., and L.S. Lee. 2010. Soil temperature and moisture effects on the persistence of synthetic androgen 17$\alpha$-trenbolone, 17$\beta$-trenbolone and trendione. Chemosphere. 79:873-879.

Kim, J.W, Y. A. Pachepsky, D.R. Shelton, and C. Coppock. 2010. Effect of streambed bacteria release on E. coli concentrations: monitoring and modeling with the modified SWAT. Ecological Modeling. 221:1592-1604.

Klein C., S. O'Connor, J. Locke, and D. Aga. 2008. Sample preparation and analysis of solid-bound pharmaceuticals. p. 81-100. *In* D.S. Aga (ed.) Fate and transport of pharmaceuticals in the environment and water treatment systems. 1st ed. CRC Press, Boca Raton, FL.

Kolodziej, E.P, T. Harter, and D.L. Sedlak. 2004. Dairy wastewater, aquaculture, and spawning fish as sources of steroid hormones in the aquatic environment. Environmental Science & Technology. 38(23):6377-6384.

Kolodziej, E.P., and D.L. Sedlak. 2007. Rangeland grazing as a source of steroid hormones to surface waters. Environmental Science & Technology. 41:3514-3520.

Kolpin, D.W., E.T. Furlong, M.T. Meyer, E.M. Thurman, S.D. Zaugg, L.B. Barber, and H.T. Buxton. 2002. Pharmaceuticals, hormones, and other organic wastewater contaminants in U.S. streams, 1999-2000: a national reconnaissance. Environmental Science & Technology. 36(6):1202-1211.

Koyuncu, I., O.A. Arikan, M.R. Wiesner, and C. Rice. 2008. Removal of hormones and antibiotics by nanofiltration membranes. Journal of Membrane Science. 309:94-101.

Kumar, K., S.C. Gupta, Y. Chander, and A.K. Singh. 2005. Antibiotic use in agriculture and its impact on the terrestrial environment. Advances in Agronomy. 87:1-54.

Kümmerer, K., 2009a. Antibiotics in the aquatic environment- a review- part I. Chemosphere. 75:417-434. http://jlakes.org/web/Antibiotics-aquatic-environment-C2009.pdf

Kümmerer, K., 2009b. Antibiotics in the aquatic environment- a review- part II. Chemosphere. 75: 435-441. http://www.jlakes.org/web/Antibiotics-aquatic-environment-PART2-C2009.pdf

Laird, A.R., V. Ibarra, G. Ruiz-Palacios, M.L. Guerrero, R.I. Glass, and J.R. Gentsch. 2003. Unexpected detection of animal VP7 genes among common rotavirus strains isolated from children in Mexico. Journal of Clinical Microbiology. 41(9):4400-4403.

Lance, J.C., and C.P. Gerba. 1984. Effect of ionic composition of suspending solution on virus adsorption by a soil column. Applied and Environmental Microbiology. 47:484-488.

Landry, E.F., J.M. Vaughn, M.Z. Thomas, and C.A. Beckwith. 1979. Adsorption of enteroviruses to soil cores and their subsequent elution by artificial rainwater. Applied and Environmental Microbiology. 38:680-687.

Lange, I.G., A. Daxenberger, B. Schiffer, H. Witters, D. Ibarreta, and H. Meyer. 2002. Sex hormones originating from different livestock production systems: fate and potential disrupting activity in the environment. Analytica Chimica Acta 473:27-37.

LeChevallier, M.W., W.D. Norton, and R.G. Lee. 1991. Occurence of *giardia* and *cryptosporidium* spp. in surface wáter supplies. Applied and Environmental Microbiology. 57(9):2610-2616.

Lee, L.S., N. Carmosini, S.S. Sassman, H.M. Dion, and M.S. Sepúlveda. 2007. Agricultural contributions of antimicrobials and hormones on soil and water quality. Advances in Argronomy.93:1-68.

Levy S.B., G. Fitzgerald, and A. Macone. 1976. Changes in intestinal flora of farm personnel after introduction of a tetracycline-supplemented feed on a farm. New England Journal of Medicine. 295:583-588.

Levy, S.B., and B. Marshall. 2004. Antibacterial resistance worldwide: causes, challenges and responses. Nature Medicine Supplement. 10(12):S122-S129.

Libby, A., and P.J. Schaible. 1955. Observations on growth responses to antibiotics and arsenic acids in poultry feeds. Science 121:733-734.

Loglisci, R. 2010. New FDA numbers reveal food animals consume lion's share of antibiotics. Johns Hopkins Healthy Monday Project, Center for a Livable Future. http://www.livablefutureblog.com/2010/12/new-fda-numbers-reveal-food-animals-consume-lion%e2%80%99s-share-of-antibiotics. (Verified February, 2012)

Lopez, C.B., E.B. Jewett, Q. Dortch, B.T. Walton, and H.K. Hudnell. 2008. Scientific assessment of freshwater harmful algal blooms: interagency working group on harmful algal blooms. Hypoxia and Human Health of the Joint Subcommittee on Ocean Science and Technology. Washington, DC. http://www.whitehouse.gov/sites/default/files/microsites/ostp/frshh2o0708.pdf. (Verified December, 2011).

López-Peñalver, J.J., M. Sánchez-Polo, C.V. Gómez-Pacheco, and J. Rivera-Utrilla. 2010. Photodegradation of tetracyclines in aqueous solution by using UV and $UV/H_2O_2$ oxidation processes. Journal of Chemical Technology and Biotechnology. 85(10):1325-1333.

Lorenzen, A., J.G. Hendel, K.L. Conn, S. Bittman, A.B. Kwabiah, G. Lazarovitz, D. Masse, T.A. McAllister, and E. Topp. 2004. Survey of hormone activities in municipal biosolids and animal manures. Environmental Toxicology. 19:216-225.

MacDonald, J.M. 2008. The economic organization of U.S. broiler production. Economic Information Bulletin Number 38. USDA, Economic Research Service. http://www.ers.usda.gov/publications/eib38/eib38.pdf. (Verified January, 2012).

MacDonald, J.M., and W.D. McBride. 2009. The transformation of U.S. livestock agriculture: scale, efficiency, and risks. Economic Information Bulletin Number 43. USDA, Economic Research Service. http://www.ers.usda.gov/Publications/EIB43/EIB43.pdf. (Verified December, 2011).

MacKenzie, W.R., N.J. Hoxie, M.E. Proctor, M.S. Gradus, K.A. Blair, K.A., D.E. Peterson, J.J. Kazmerczak, D.G. Addiss, K.R. Fox, J.B. Rose, and J.P. Davis. 1994. A massive outbreak in Milwaukee of *Cryptosporidium* infection transmitted through the public water supply. New England Journal of Medicine. 331:161-167.

MacMillan, J.R., R. Schnick, and G. Fornshell. 2003. Volume of antibiotics sold (2001 and 2002) in U.S. domestic aquaculture industry. The National Aquaculture Association. http://www.thenaa.net/downloads/AETF_Antibiotic_use_white_paper_6.11.03.pdf. (Verified December, 2011).

Mallin, M.A., and L.B. Cahoon. 2003. Industrialized animal production- a major source of nutrient and microbial pollution to aquatic ecosystems. Population and Environment. 24(5):369-385.

Marin, B., M. Klingberg, and M. Melkonian. 1998. Phylogenetic relationships among the Cryptophyta: analyses of nuclear-encoded ssu rRNA sequences support the monophyly of plastid containing lineages. Protist. 149:264–276.

Marks, R. 2001. Cesspools of shame: how factory farm lagoons and sprayfields threaten environmental and public health. Natural Resources Defense Council and the Clean Water Network. Washington, DC. http://www.nrdc.org/water/pollution/cesspools/cesspools.pdf. (Verified December, 2011).

Mattison, K., A. Shukla, A. Cook, F. Pollari, R. Friendship, D. Kelton, S. Bidawid, and J.M. Farber. 2007. Human noroviruses in swine and cattle. Emerging Infectious Diseases. 13(8):1184-1188.

Mawdsley, J.L., A.E. Brooks, R.J. Merry, and B.F. Pain. 1996. Use of a novel soil tilting table apparatus to demonstrate the horizontal and vertical movement of the protozoan pathogen *Cryptosporidium parvum* in soil. Biology and Fertility of Soils. 23(2):215-220.

McEwen, S., and P.J. Fedorka-Cray. 2002. Antimicrobial use and resistance in animals. Clinical Infectious Diseases. 34(Suppl 3):S93-106.

McGuffey, R.K., L.F. Richardson, and J.I.D. Wilkinson. 2001. Ionophores for dairy cattle: current status and future outlook. Journal of Dairy Science. 84(E. Suppl.):E194-E203.

Mellon, M., C. Benbrook, K.L. Benbrook. 2001. Hogging it, estimates of antimicrobial abuse in livestock. Union of Concerned Scientists. http://www.ucsusa.org/food_and_agriculture/science_and_impacts/impacts_industrial_agriculture/hogging-it-estimates-of.html. (Verified December, 2011).

Meng, X.J., R.H. Purcell, P.G. Halbur, J.R. Lehman, D.M. Webb, T.S. Tsareva, J.S. Haynes, B.J. Thacker, and S.U. Emerson. 1997. A novel virus in swine is closely related to the human hepatitis E virus. Proceedings of the National Academy of Sciences of the United States of America. 94:9860-9865.

Merchant, J.A., A.L. Naleway, E.R. Svendsen, K.M. Kelly, L.F. Burmeister, A.M. Stromquist, C.D. Taylor, P.S. Thorne, S.J. Reynolds, W.T. Sanderson, and E.A. Chrischilles. 2005. Asthma and farm exposures in a cohort of rural Iowa children. Environmental Health Perspectives. 113(3):350-356.

Mills, A.L., J.S. Herman, G.M. Hornberger, and T.H. DeJesus. 1994. Effect of solution ionic strength and iron coatings on mineral grains on the sorption of bacterial cells to quartz sand. Applied and Environmental Microbiology. 60:3300-3306.

Midwest Planning Service – Livestock Waste Subcommittee. 1985. Livestock Waste Facilities Handbook. Midwest Planning Service Report MWPS-18, 2nd ed. Ames, Iowa: Iowa State University.

Mirabelli, M.C., S. Wing, S.W. Marshall, and T.C. Wilcosky. 2006. Asthma symptoms among adolescents who attend public schools that are located near confined swine feeding operations. Pediatrics. 118:e66-75.

Moe, C.L., M.D. Sobsey, P.W. Stewart, and D. Crawford-Brown. 1999. Estimating the risk of human calicivirus infection from drinking water. Poster P4-6, First International Calicivirus Workshop. March, 1999, Atlanta, GA.

Mohammed HO, K Stipetic, P L McDonough, R N Gonzalez RN, D V Nydam, and E R Atwill.. 2009. Identification of potential on-farm sources of Listeria monocytogenes in herds of dairy cattle. Am J Vet Res. 70(3):383-8.

Muirhead, R., R. Collins, and P. Bremer. 2006. Numbers and transported state of *Escherichia coli* in runoff direct from fresh cowpats under simulated rainfall. Letters in Applied Microbiology. 42:83–87.

Mulla, D.J., A. Selely, A. Birr, J. Perry, B. Vondracek, E. Bean, E. Macbeth, S. Goyal, B. Wheeler, C. Alexander, G. Randall, G. Sands, and J. Linn. 1999. Generic environmental impact statement on animal agriculture: a summary of the literature related to the effects of animal agriculture on water resources. Minnesota Environmental Quality Board, Minnesota Department of Agriculture.

National Antimicrobial Resistance Monitoring System (NARMS). 2009. NARMS retail meat annual report 2009. http://www.fda.gov/AnimalVeterinary/SafetyHealth/AntimicrobialResistance/NationalAnti microbialResistanceMonitoringSystem/ucm257561.htm. (Verified December, 2011).

National Research Council (NRC). 1993. Soil and water quality: an agenda for agriculture. Long-Range Soil and Water Conservation Policy, NRC. National Academy of Sciences: Washington, DC.

NRC. 1999. The use of drugs in food animals: benefits and risks. Food and Nutrition Board, Institute of Medicine. National Academy Press, Washington, DC.

Natural Resources Conservation Service/U.S. Department of Agriculture (NRCS/USDA) 2012. Nutrient Management Technical Note No. 9. Introduction to Waterborne Pathogens in Agricultural Watersheds. http://www.google.com/url?sa=t&rct=j&q=Nutrient+Management+Technical+Note +No.+9.+Introduction+to+Waterborne+Pathogens+in+Agricultural+Watersheds&source=web&c d=1&ved=0CDQQFjAA&url=http%3A%2F%2Fdhs.wifss.ucdavis.edu%2Fheadcontent%2Fnewsle tter%2F_12nov%2FNCRC-WaterPathTechNote-Sep-2012.pdf&ei=4O7sUKicPM-- 0QGQhIFQ&usg=AFQjCNFkV6IMmiSBNILDTqZpSwcNETuVfg&bvm=bv.1357316858,d.dmQ (Verfied January 2013).

Nelson, J.M., T.M. Chiller, J.H. Powers, and F.J. Angulo. 2007. Fluoroquinolone-resistant *Campylobacter* species and the withdrawal of fluoroquinolones from use in poultry: a public health success story. Food Safety. 44:977-980.

Neumeister, L. 2012. NY judge: FDA should act on animal antibiotics. Business Week. http://www.businessweek.com/ap/2012-06/D9V77NPO1.htm. (Verified July, 2012).

New York Department of Environmental Conservation (NYSDEC). 2007. DEC reports: progress since Marks dairy spill. August 9, 2007. http://www.dec.ny.gov/press/36942.html. (Verified December, 2011).

Nghiem, L.D., A.I. Schäfer, and M. Elimelech. 2004. Removal of natural hormones by nanofiltration membranes: measurement, modeling, and mechanisms. Environmental Science & Technology. 38:1888-1896.

Nutrient Innovations Task Group (NITG). 2009. An urgent call to action- report of the state-EPA nutrient innovations task group. August, 2009.

Ohio Environmental Protection Agency (OEPA). 2007. Beaver Creek and Grand Lake St. Marys watershed TMDL report. OEPA Fact Sheet. October 2007. http://www.epa.ohio.gov/portals/35/tmdl/GLSMfactsheet%20Final%20Oct07.pdf. (Verified December, 2011).

OEPA. 2009. Efforts to improve water quality in Grand Lake St. Marys. http://www.epa.ohio.gov/portals/47/citizen/efforts_to_improve_water_quality_in_grand_lake_st_marys_june2009.pdf

OEPA. 2011. Grand Lake St. Marys toxic algae. May 2011. http://www.epa.ohio.gov/pic/glsm_algae.aspx. (Verified December, 2011).

Olson, M.E. 2001. Human and animal pathogens in manure. Presented at the Livestock Options for the Future Conference. June 25-27, 2001. Winnipeg, Manitoba, Canada.

Orlando, E.F., A.S. Kolok, G.A. Binzcik, J.L. Gates, M.K. Horton, C.S. Lambright, L.E. Gray, A.M. Soto, and L.J. Guillette. 2004. Endocrine-disrupting effects of cattle feedlot effluent on an aquatic sentinel species, the fathead minnow. Environmental Health Perspectives. 112(2):353-358.

Osterburg, D., and D. Wallinga. 2004. Addressing externalities from swine production to reduce public health and environmental impact. American Journal of Public Health. 94:1703-1708.

Pachepsky, Y., A. Sadeghi, S. Bradford, D. Shelton, A. Guber, and T. Dao. 2006. Transport and fate of manure-borne pathogens: modeling perspective. Agricultural Water Management. 86:81- 92.

Pachepsky, Y., D. R. Shelton, J.E.T. McLain, J. Patel, R. Mandrell. 2011. Irrigation Waters as a Source of Pathogenic Microorganisms in Produce. A Review. Advances in Agronomy, 113:73-138.

Pachepsky, Y., J. Morrow, A. Guber, D. Shelton, R. Rowland, G. Davies. 2012. Effect of biofilm in irrigation pipes on microbial quality of irrigation water. Letters in Applied Microbiology, 54: 217-224.

Palme, R., P. Fischer, H. Schildofer, and M.N. Ismail. 1996. Excretion of infused 14C-steroid hormones via faeces and urine in domestic livestock. Animal Reproduction Science. 43:43-63.

Pappas, E., R. Kanwar, J. Baker, J. Lorimor, and S. Mickelson. 2008. Fecal indicator bacteria in subsurface drain water following swine manure application. American Society of Agricultural and Biological Engineers. 51(5):1567-1573.

Payment, P. 1989. Presence of human and animal viruses in surface and ground waters, Water Science and Technology. 21:283-285.

Peng, M.M., L. Xiao, A.R. Freeman, M.J. Arrowood, A.A. Escalante, A.C. Weltman, C.S.L. Ong, W.R. Mac Kenzie, A.A. Lal, and C.B. Beard. 1997. Genetic polymorphism among Cryptosporidium parvum isolates: evidence of two distinct human transmission cycles. Emerging Infectious Diseases. 3:567-573.

Perdek, J.M., R.D. Arnone, M.K. Stinson, and M.E. Tuccillo. 2003. Managing pathogens in the urban watershed. EPA-600-R-03-111. USEPA, Office of Research and Development, National Risk Management Research Laboratory. Cincinnati, OH.

Pérez, S., and D. Barceló. 2008. Advances in the analysis of pharmaceuticals in the aquatic environment. p. 53-80. *In* D.S. Aga (ed.) Fate and transport of pharmaceuticals in the environment and water treatment systems. 1st ed. CRC Press, Boca Raton, FL.

Pesaro, F., I. Sorg, and A. Metzler. 1995. In situ inactivation of animal viruses and a coliphage in nonaerated liquid and semiliquid animal wastes. Applied and Environmental Microbiology. 61(1):92-97.

Pew Commission on Industrial Farm Animal Production (PCIFAP). 2008. Putting meat on the table: industrial farm animal production in America. The Pew Charitable Trusts and Johns Hopkins Bloomberg School of Public Health. http://www.pewtrusts.org/our_work_report_detail.aspx?id=38442. (Verified December, 2011).

Plourde, J.R., H. Hafez-Zaden, and J.P. Lemoine. 1974. Biotransformation of progesterone by spores and vegetative cells of micro-organisms found in the soil of Quebec. Reviews of Canadian Biology. 33:111-116.

Pokorna, J., and A. Kasal. 1990. Progesterone side-chain degradation beside hydroxylation with *Rhizopus nigricans* depends on the presence of nutrients. Journal of Steroid Biochemistry. 35:155-156.

Price, L.B., E. Johnson, R. Vailes, and E. Silbergeld. 2005. Fluoroquinolone-resistant *Campylobacter* isolates from conventional and antibiotic-free chicken products. Environmental Health Perspectives. 113(5):557-560.

Price, L.B., L.G. Lackey, R. Vailes, and E. Silbergeld. 2007. The persistence of fluoroquinolone-resistant *Campylobacter* in poultry production. Environmental Health Perspectives. 115(7):1035-1039.

Public Health Agency of Canada (PHAC). 2000. Waterborne outbreak of gastroenteritis associated with a contaminated municipal water supply, Walkerton, Ontario, May-June 2000. October 15, 2000. Canada Communicable Disease Report. Vol. 26-20. http://wvlc.uwaterloo.ca/biology447/modules/module4/HealthCanadasummary.htm. (Verified December, 2011).

PHAC. 2010. Hepatitis E virus: pathogen safety data sheet – infectious substances. http://www.phac-aspc.gc.ca/lab-bio/res/psds-ftss/hepe-eng.php. (Verified December, 2011).

Puckett, L.J. 1994. Nonpoint and point sources of nitrogen in major watersheds of the United States. Water-Resources Investigations Report 94-4001. USGS, Reston, VA.

Pyle, B.H., S.C. Broadaway, and G.A. McFeters. (1999). Sensitive detection of *Escherichia coli* O157:H7 in food and water by immunomagnetic separation and solid-phase laser cytometry. Applied and Environmental Microbiology. 65(5):1966-1972.

Raman, D.R., A.C. Layton, L.B. Moody, and J.P. Easter. 2001. Degradation of estrogens in dairy waste solids: effects of acidification and temperature. Transactions of the ASAE. 44:1881-1888.

Raman D.R., E.L. Williams, A.C. Layton, R.T. Burns, J.P. Easter, A.J. Daugherty, D. Mullen, and G.S. Sayler. 2004. Estrogen content of dairy and swine wastes. Environmental Science & Technology. 38:3567-3573.

Ramaswamy, J., S.O. Prasher, R.M. Patel, S.A. Hussain, and S.F. Barrington. 2010. The effect of composting on the degradation of a veterinary pharmaceutical. Bioresource Technology. 101:2294-2299.

Reddy, K.R., R. Khaleel, and M.R. Overcash. 1981. Behavior and transport of microbial pathogens and indicator organisms in soils treated with organic wastes. Journal of Environmental Quality. 10:255-265.

Ribaudo, M., and N. Gollehon. 2006. Animal agriculture and the environment. p. 124-133. *In* K. Wiebe and N. Gollehon (ed.) Agriculture resources and environmental indicators, 2006 edition. Economic Information Bulletin No. (EIB-16). USDA, NRCS, ERS, Washington, DC. http://www.ers.usda.gov/publications/arei/eib16/. (Verified December, 2011).

Richardson, A.J., R.A. Frankenberg, A.C. Buck, J.B. Selkon, J.S. Colbourne, J.W. Parsons, and R.T. Mayon-White. 1991. An outbreak of waterborne cryptosporidiosis in Swindon and Oxfordshire, Epidemiology and Infection. 107(3):485-95.

Rogers, S., and J. Haines. 2005. Detecting and mitigating the environmental impact of fecal pathogens originating from confined animal feeding operations: review. EPA-600-R-06-021. USEPA, Office of Research and Development, National Risk Management Research Laboratory. Cincinnati, OH.

Rogers, S. 2011. Zoonotic disease agents: livestock sources, transport pathways, and public health risks. PowerPoint Presentation for USEPA, Office of Water, Office of Science and Technology on September 7, 2011.

Rogers, S.W., M. Donnelly, L. Peed, S. Mondal, Z. Zhang, and O. Shanks. 2011. Decay of bacterial pathogens, fecal indicators, and real time quantitative PCR genetic markers in manure-amended soils. Applied Environmental Microbiology. 77(14):4839-4848.

Rose, J.B. 1997. Environmental ecology of *Cryptosporidium* and public health implications. Annual Review of Public Health. 18:135-161.

Rosen, B.H. 2000. Waterborne pathogens in agricultural watersheds. Watershed Science Institute United States. USDA, Natural Resources Conservation Service, Washington, DC.

Rosenfeldt, E.J., and K.G. Linden. 2004. Degradation of endocrine disrupting chemicals bisphenol A, ethinyl estradiol, and estradiol during UV photolysis and advanced oxidation processes. Environmental Science & Technology. 38(20):5476-5483.

Royer, M., L. Xiao, and A. Lal. 2002. Animal source identification using a *Cryptosporidium* DNA characterization technique. EPA-600-R-03-047. USEPA, Office of Research and Development, National Risk Management Research Laboratory. Cincinnati, OH.

Ruddy, B.C., D.L. Lorenz, and D.K. Mueller. 2006. County-level estimates of nutrient inputs to the land surface of the conterminous United States, 1982–2001. U.S. Geological Survey Scientific Investigations Report 2006-5012, Reston, VA.

Sahlström, L. 2003. A review of survival of pathogenic bacteria in organic waste used in biogas plants. Bioresource Technology. 87:161-166.

Sanderson, H., R.A. Brain, D.J. Johnson, C.J. Wilson, and K.R. Solomon. 2004. Toxicity classification and evaluation of four pharmaceuticals classes: antibiotics, antineoplastics, cardiovascular, and sex hormones. Toxicology. 203:27-40.

Sapkota, A.R., F.C. Curriero, K.E. Gibson, and K.J. Schwab. 2007. Antibiotic-resistant enterococci and fecal indicators in surface water and groundwater impacted by a concentrated swine feeding operation. Environmental Health Perspectives. 115(7):1040-1045.

Sapkota, A.R., R.M. Hulet, G. Zhang, P. McDermott, E.L. Kinney, K.J. Schwab, and S.W. Joseph. 2011. Lower prevalence of antibiotic-resistant enterococci on U.S. conventional poultry farms that transitioned to organic practices. Environmental Health Perspectives. 119(11):1622-1628.

Sarmah, A.K., M.T. Meyer, A. Boxall. 2006. A global perspective on the use, sales, exposure pathways, occurrence, fate and effects of veterinary antibiotics (VAs) in the environment. Chemosphere. 65:725-759.

Sayah, R.S., J.B. Kaneene, Y. Johnson, and R. Miller. 2005. Patterns of antimicrobial resistance observed in *Escherichia coli* isolates obtained from domestic- and wild-animal fecal samples, human septage, and surface water. Applied and Environmental Microbiology. 71(3):1394-1404.

Schenkler, G., W. Muller, and P. Glatzel. 1998. Continuing studies on the stability of sex steroids in the feces of cows over 12 weeks. Berliner Münchener tierärztliche Wochenschrift. 111:248-252.

Schiffer, B., A. Daxenberger, K. Meyer, H.H.D. Meyer. 2001. The fate of trenbolone acetate and melengestrol acetate after application as growth promoters in cattle: environmental studies. Environmental Health Perspectives. 109(11):1145-1150.

Scholl, M.A., and R.W. Harvey. 1992. Laboratory investigations on the role of sediment surface and groundwater chemistry in transport of bacteria through a contaminated sandy aquifer. Environmental Science & Technology. 26(7):1410-1417.

Schumacher, J.G. 2003. Survival, transport, and sources of fecal bacteria in streams and survival in land-applied poultry litter in the upper Shoal Creek basin, southwestern Missouri, 2001–2002. USGS Water-Resources Investigations Report 03–4243.

Schwarzenberger, F., E. Möstl , R. Palme, and E. Bamberg. 1996. Faecal steroid analysis for non-invasive monitoring of reproductive status in farm, wild, and zoo animals. Animal Reproduction Science. 42:515-526.

Scott, C., H. Smith, and A. Gibbs. 1994. Excretion of *Cryptosporidium parvum* oocysts by a herd of beef suckler cows. The Veterinary Record. 134:172.

Sellin, M.K., D.D. Snow, S.T. Gustafson, G.E. Erickson, and A.S. Kolok. 2009 The endocrine activity of beef cattle wastes: do growth-promoting steroids make a difference? Aquatic Toxicology. 92(4):221–227.

Sengeløv, G., Y. Agersø, B. Halling-Sørensen, S.B. Baloda, J.S. Anderson, and L.B. Jensen. 2003. Bacterial antibiotic resistance levels in Danish farmland as a result of treatment with pig manure slurry. Environment International. 28(7):587-595.

Shore, L.S., E. Harel-Markowitz, M. Gurevich, and M. Shemesh. 1993. Factors affecting the concentration of testosterone in poultry litter. Journal of Environmental Science and Health. A28:1737-1749.

Shore, L.S., D.L. Correll, and P.K. Chakraborty. 1995. Relationship of fertilization with chicken manure and concentration of estrogens in small streams. p. 155-162. *In* K. Steele (ed.) Animal Waste and the Land-Water Interface. Lewis Publishing, Boca Raton, FL.

Shore, L.S., and M. Shemesh. 2003. Naturally produced steroid hormones and their release into the environment. Pure and Applied Chemistry. 75:1859-1871.

Shore, L.S. 2009. Steroid hormones generated by CAFOs. p. 13-21. *In* L.S. Shore and A. Pruden (ed.) Hormones and pharmaceuticals generated by concentrated animal feeding operations. Emerging Topics in Ecotoxicology. Vol. 1.

Simon R., and J. Makarewicz. 2009. Impacts of manure management practices on stream microbial loading into Conesus Lake, NY. Journal of Great Lakes Research 35:66–75.

Smith, H.V., W.J. Patterson, R. Hardie, L.A. Greene, C. Benton, W. Tulloch, R.A. Gilmour, R.W.A. Girdwood, J.C.M. Sharp, and G.I. Forbes. 1989. An outbreak of waterborne cryptosporidiosis caused by post-treatment contamination. Epidemiology and Infection. 103:703-715.

Snyder, S.A., P. Westerhoff, Y. Yoon, and D.L. Sedlak. 2003. Pharmaceuticals, personal care products, and endocrine disruptors in water: implications for the water industry. Environmental Engineering Science. 20(5):449-469.

Snyder, S.A., B. Vanderford, R. Trenholm, J. Holady, and D. Rexing. 2005. Occurrence of EDCs and pharmaceuticals in U.S. drinking waters. Presented at the American Water Works Association Water Quality Technology Conference. November 9, 2005. Quebec City, Quebec, Canada.

Snyder, S.A., S. Adham, A.M. Redding, F.S. Cannon, J. DeCarolis, J. Oppenheimer, E.C. Wert, and Y. Yoon. 2006. Role of membranes and activated carbon in the removal of endocrine disruptors and pharmaceuticals. Desalination. 202:156-181.

Snyder, S.A., H. Lei, and E.C. Wert. 2008. Removal of endocrine disruptors and pharmaceuticals during water treatment. p. 229-259. *In* D.S. Aga (ed.) Fate and transport of pharmaceuticals in the environment and water treatment systems. 1st ed. CRC Press, Boca Raton, FL.

Sobsey, M.D., L.A. Khatib, V.R. Hill, E. Alocilja, and S. Pillai. 2006. Pathogens in animal wastes and the impacts of waste management practices on their survival, transport and fate. p. 609-666. *In* J. M. Rice, D. F. Caldwell, and F. J. Humenik (ed.) Animal Agriculture and the Environment: National Center for Manure and Animal Waste Management White Papers. American Society of Agricultural and Biological Engineers. St. Joseph, Michigan.

Soller, J., M. Schoen, T. Bartrand, J. Ravenscroft, and N. Ashbolt. 2010. Estimated human health risks from exposure to recreational waters impacted by human and non-human sources of faecal contamination. Water Research. 44(16): 4674-4691.

Solo-Gabriele, H., and S. Neumeister. 1996. U.S. outbreaks of cryptosporidiosis. Journal American Water Works Association. 88(9): 76-86.

Soupir, M., and S. Mostaghimi. 2011. *Escherichia coli* and enterococci attachment to particles in runoff from highly and sparsely vegetated grassland. Water Air Soil Pollution. 216:167-178.

Stackelberg, P.E., J. Gibs, E.T. Furlong, M.T. Meyer, S.D. Zaugg, and R.L. Lippincott. 2007. Efficiency of conventional drinking-water-treatment processes in removal of pharmaceuticals and other organic compounds. Science of the Total Environment. 377:255-272.

Steinfeld, H., P. Gerber, W. Wassenaar, V. Castel, M. Rosales, and C. de Haan. 2006. Livestock's long shadow- environmental issues and options. The Livestock, Environment and Development (LEAD) Initiative and the Food and Agriculture Organization of the United Nations, Rome, Italy.

Suter, E., A.R. Juhl, and G. D. O'Mullan. 2011. Particle association of *Enterococcus* and total bacteria in the Lower Hudson River Estuary, USA. Journal of Water Resource and Protection. 3:715-725.

Swartz, M.N. 1989. Human health risks with the subtherapeutic use of penicillin or tetracyclines in animal feed. Committee to Study the Human Health Effects of Subtherapeutic Antibiotic Use in Animal Feeds, Division of Medical Sciences, Assembly of Life Sciences, National Research Council, Washington, DC.

Swartz, M.N. 2002. Human diseases caused by foodborne pathogens of animal origin. Clinical Infectious Diseases. 34(Suppl 3):S111-S122.

Swerdlow, D.L., B.A. Woodruff, R.C. Brady, P.M. Griffin, S. Tippen, H.D. Donnell, E. Geldreich, B.J. Payne, A. Meyer, J.G. Wells, K.D. Greene, M. Bright, N.H. Bean, and P.A. Blake. 1992. A waterborne outbreak in Missouri of *Escherichia coli* O157:H7 associated with bloody diarrhea and death. Annals of Internal Medicine. 117(10):812-819.

Szenci, O., R. Palme, M.A.M. Taverne, J. Varga, N. Meersma, and E. Wissink. 1997. Evaluation of false ultrasonographic pregnancy diagnosis in sows by measuring the concentration of unconjugated estrogens in feces. Theriogenology. 48:873-882.

Szogi, A.A., P.J. Bauer, and M.B. Vanotti. 2010. Fertilizer effectiveness of phosphorus recovered from broiler litter. Agronomy Journal. 102(2): 723-727.

Tanner, A.C. 2000. Antimicrobial drug use in poultry. p. 637-655. *In* J.F. Prescott, J.D. Baggot, R.D. Walker (ed.) Antimicrobial Therapy in Veterinary Medicine. 3rd ed. Iowa State University Press, Ames, IA.

Tetra Tech, Inc. 2002. State compendium: programs and regulatory activities related to animal feeding operations. USEPA, Office of Wastewater Management, Washington, DC. http://www.epa.gov/npdes/pubs/statecom.pdf. (Verified December, 2011).

Thurston-Enriquez, J., J. Gilley, and B. Eghbal, B. 2005. Microbial quality of runoff following land application of cattle manure and swine slurry. Journal of Water and Health. 3.2: 157-171.

Tyrrel, S.F., and J.N. Quinton. 2003. Overland flow transport of pathogens from agricultural land receiving faecal wastes. Journal of Applied Microbiology. 94:87S-93S.

Unc, A., and M.J. Goss. 2003. Movement of faecal bacteria through the vadose zone. Water, Air, and Soil Pollution. 149:327-337.

Unc, A., and M.J. Goss. 2004. Transport of bacteria from manure and protection of water resources. Applied Soil Ecology. 25:1-18.

United States Congress, Office of Technology Assessment (OTA). 1995. Impacts of antibiotic-resistant bacteria. OTA-H-629. U.S. Government Printing Office, Washington, DC.

United States Department of Agriculture (USDA). 1999. 1997 Census of agriculture. AC97-A-51. Volume 1. USDA, National Agricultural Statistics Service, Geographic Area Series, Washington, DC.

USDA. 2000. Part I: baseline reference of feedlot management practices, 1999. Report No. #N327.0500. USDA, APHIS, VS, CEAH, National Animal Health Monitoring System, Fort Collins, CO. http://www.aphis.usda.gov/animal_health/nahms/feedlot/downloads/feedlot99/Feedlot99_dr_PartI.pdf. (Verified December, 2011).

USDA. 2002a. Part III: reference of swine health and environmental management in the United States, 2000. Report No. #N361.0902. USDA, APHIS, VS, CEAH, National Animal Health Monitoring System, Fort Collins,

CO. http://www.aphis.usda.gov/animal_health/nahms/swine/downloads/swine2000/Swine2000_d
r_PartIII.pdf. (Verified December, 2011).

USDA. 2002b. Preventive practices in swine: administration of iron and antibiotics. Info Sheet, March 2002.
USDA, APHIS, VS, CEAH, National Animal Health Monitoring System, Fort Collins,
CO. http://www.aphis.usda.gov/animal_health/nahms/swine/downloads/swine2000/Swine2000_is
_Iron.pdf. (Verified December, 2011).

USDA. 2006. Census of aquaculture (2005). Volume 3, Special Studies Part 2. AC-02-SP-2. USDA, National
Agricultural Statistics Service, Washington, DC.

USDA. 2007a Swine 2006: Part I: reference of swine health and management practices in the United States
2006. USDA, APHIS, VS, National Animal Health Monitoring System, Fort Collins,
CO. http://www.aphis.usda.gov/animal_health/nahms/swine/downloads/swine2006/Swine2006_d
r_PartI.pdf. (Verified December, 2011).

USDA. 2007b Swine 2006: Part II: reference of swine health and management practices in the United States
2006. USDA, APHIS, VS, National Animal Health Monitoring System, Fort Collins,
CO. http://www.aphis.usda.gov/animal_health/nahms/swine/downloads/swine2006/Swine2006_d
r_PartII.pdf. (Verified December, 2011).

USDA. 2007c. Composting manure- what's going on in the dark? Manure Management Information Sheet,
Number 1, May 2007. USDA, Natural Resources Conservation
Service. http://www.or.nrcs.usda.gov/technical/engineering/environmental_engineering/Compost_
Netmeeting/Composting_Manure_Info_Sheet.pdf. (Verified December, 2011).

USDA. 2008a. Antibiotic use on U.S. dairy operations, 2002 and 2007. Info Sheet, October 2008. USDA,
APHIS, VS, CEAH, National Animal Health Monitoring System, Fort Collins,
CO. http://www.aphis.usda.gov/animal_health/nahms/dairy/downloads/dairy07/Dairy07_is_Anti
bioticUse.pdf. (Verified December, 2011).

USDA. 2008b. Disease prevention, treatment practices, and antibiotic administration techniques on U.S.
swine sites. Info Sheet, March 2008. USDA, APHIS, VS, CEAH, National Animal Health
Monitoring System, Fort Collins,
CO. http://www.aphis.usda.gov/animal_health/nahms/swine/downloads/swine2006/Swine2006_is
_DisPrev.pdf. (Verified December, 2011).

USDA. 2008c. *Campylobacter* on U.S. swine sites- antimicrobial susceptibility. Info Sheet, December 2008.
USDA, APHIS, VS, CEAH, National Animal Health Monitoring System, Fort Collins,
CO. http://www.aphis.usda.gov/animal_health/nahms/swine/downloads/swine2006/Swine2006_is
_campy.pdf. (Verified December, 2011).

USDA. 2009a. 2007 Census of agriculture. AC-07-A-51. Volume 1. USDA, National Agricultural Statistics
Service. Geographic Area Series, Washington,
DC. http://www.agcensus.usda.gov/Publications/2007/Full_Report/usv1.pdf. (Verified December,
2011).

USDA. 2009b. Dairy 2007: Part I: reference of dairy cattle health and management in the United States, 2007. USDA, APHIS, VS, National Animal Health Monitoring System, Fort Collins, CO. http://www.aphis.usda.gov/animal_health/nahms/dairy/downloads/dairy07/Dairy07_dr_Part I.pdf. (Verified December, 2011).

USDA. 2009c. Reproduction practices on U.S. dairy operations, 2007. Info Sheet, February 2009. USDA, APHIS, VS, CEAH, National Animal Health Monitoring System, Fort Collins, CO. http://www.aphis.usda.gov/animal_health/nahms/dairy/downloads/dairy07/Dairy07_is_Repr odPrac.pdf. (Verified December, 2011).

USDA. 2009d. Dairy 2007: Part IV: reference of dairy cattle health and management practices in the United States, 2007. USDA, APHIS, VS, National Animal Health Monitoring System, Fort Collins, CO. http://www.aphis.usda.gov/animal_health/nahms/dairy/downloads/dairy07/Dairy07_dr_Part IV.pdf. (Verified February, 2012).

USDA. 2009e. *Salmonella* on U.S. beef cow-calf operations, 2007-08. Info Sheet, June 2009. USDA, APHIS, VS, CEAH, National Animal Health Monitoring System, Fort Collins, CO. http://www.aphis.usda.gov/animal_health/nahms/beefcowcalf/downloads/beef0708/Beef070 8_is_Salmonella.pdf. (Verified December, 2011).

USDA. 2009f. *Salmonella* and *Campylobacter* on U.S. dairy operations, 1996-2007. Info Sheet, July 2009. USDA, APHIS, VS, CEAH, National Animal Health Monitoring System, Fort Collins, CO. http://www.aphis.usda.gov/animal_health/nahms/dairy/downloads/dairy07/Dairy07_is_SalC ampy.pdf. (Verified December, 2011).

USDA. 2009g. *Salmonella* on U.S. swine sites- prevalence and antimicrobial susceptibility. Info Sheet, January 2009. USDA, APHIS, VS, CEAH, National Animal Health Monitoring System, Fort Collins, CO. http://www.aphis.usda.gov/animal_health/nahms/swine/downloads/swine2006/Swine2006_is _salmonella.pdf. (Verified December, 2011).

USDA. 2009h. *Escherichia coli* on U.S. swine sites- antimicrobial susceptibility. Info Sheet, January 2009. USDA, APHIS, VS, CEAH, National Animal Health Monitoring System, Fort Collins, CO. http://www.aphis.usda.gov/animal_health/nahms/swine/downloads/swine2006/Swine2006_is _ecoli.pdf. (Verified December, 2011).

USDA. 2009i. *Campylobacter* on U.S. beef cow-calf operations, 2007-08. Info Sheet, June 2009. USDA, APHIS, VS, CEAH, National Animal Health Monitoring System, Fort Collins, CO. http://www.aphis.usda.gov/animal_health/nahms/beefcowcalf/index.shtml. (Verified December, 2011).

USDA. 2009j. Composting manure: small scale solutions for your farm. USDA, Natural Resources Conservation Service. http://www.ri.nrcs.usda.gov/technical/PDF/Small_Scale_Farm_Practices/Composting_Ma nure.pdf. (Verified December, 2011).

USDA. 2010a. Beef 2007-08: Prevalence and control of bovine viral diarrhea virus on U.S. cow-calf operations, 2007-08. USDA, APHIS, VS, National Animal Health Monitoring System, Fort Collins,

CO. http://www.aphis.usda.gov/animal_health/nahms/beefcowcalf/downloads/beef0708/Beef070
8_ir_BVD.pdf. (Verified December, 2011).

USDA. 2010b. Beef 2007-08 Part IV: reference of beef cow-calf management practices in the United States,
2007-08. USDA, APHIS, VS, National Animal Health Monitoring System, Fort Collins,
CO. http://www.aphis.usda.gov/animal_health/nahms/beefcowcalf/downloads/beef0708/Beef070
8_dr_PartIV.pdf. (Verified December, 2011).

USDA. 2010c. Catfish 2010, Part I: reference of catfish health and production practices in the United States,
2009. USDA, APHIS, VS, National Animal Health Monitoring System, Fort Collins,
CO. http://www.aphis.usda.gov/animal_health/nahms/aquaculture/downloads/catfish10/Cat10_d
r_PartI.pdf. (Verified December, 2011).

USDA. 2011a. Highlights of health and management practices on breeder chicken farms in the United States,
2010. Info Sheet, November 2011. USDA, APHIS, VS, CEAH. National Animal Health Monitoring
System, Fort Collins,
CO. http://www.aphis.usda.gov/animal_health/nahms/poultry/downloads/poultry10/Poultry10_is
_Breeder_highlights.pdf. (Verified February, 2012).

USDA. 2011b. Losses caused by enteric septicemia of catfish (ESC) 2002-09. Info Sheet, August 2011.
USDA, APHIS, VS, CEAH, National Animal Health Monitoring System, Fort Collins,
CO. http://www.aphis.usda.gov/animal_health/nahms/aquaculture/downloads/catfish10/Cat10_is
_ESC.pdf. (Verified December, 2011).

USDA. 2011c. Anaerobic digester, controlled temperature. USDA, Natural Resources Conservation
Service. http://www.ny.nrcs.usda.gov/technical/practices/pc366.html. (Verified December, 2011).

United States Environmental Protection Agency (USEPA). 2001. Source water protection practices bulletin
managing livestock, poultry, and horse waste to prevent contamination of drinking water. EPA-916-
F-01-026. USEPA, Office of Water, Washington,
DC. http://www.epa.gov/safewater/sourcewater/pubs/fs_swpp_livestock.pdf. (Verified February,
2012).

USEPA. 2002a. Environmental and economic benefit analysis of final revisions to the national pollutant
discharge elimination system regulation and the effluent guidelines for concentrated animal feeding
operations. EPA-821-R-03-003. USEPA, Office of Water, Washington,
DC. http://cfpub.epa.gov/npdes/afo/aforule.cfm. (Verified December, 2011).

USEPA. 2002b. Development document for the final revisions to the national pollutant discharge elimination
system regulation and the effluent guidelines for concentrated animal feeding operations. EPA-821-
R-03-001. USEPA, Office of Water, Washington,
D.C. http://cfpub.epa.gov/npdes/afo/aforule.cfm. (Verified December, 2011).

USEPA. 2004a. Risk management evaluation for concentrated animal feeding operations. EPA-600-R-04-042.
USEPA, Office of Research and Development, National Risk Management Research Laboratory,
Cincinnati, OH.

USEPA. 2004b. Drinking water treatment. USEPA 816-F-04-034. USEPA, Office of
Water. http://water.epa.gov/lawsregs/guidance/sdwa/upload/2009_08_28_sdwa_fs_30ann_treatm
ent_web.pdf. (Verified December, 2011).

USEPA. 2007. Effect of treatment on nutrient availability. USEPA, Office of Water, Office of Ground Water
and Drinking Water, Total Coliform Rule Issue Paper, Washington,
DC. http://www.epa.gov/ogwdw/disinfection/tcr/pdfs/issuepaper_tcr_treatment-nutrients.pdf

USEPA. 2009a. Ag 101 beef production: manure management system. Updated September 10,
2009. http://www.epa.gov/oecaagct/ag101/beefmanure.html. (Verified December, 2011).

USEPA. 2009b. National water quality inventory: report to Congress 2004 Reporting Cycle. EPA-841-R-08-
001. USEPA, Office of Water, Washington,
DC. http://water.epa.gov/lawsregs/guidance/cwa/305b/2004report_index.cfm. (Verified
December, 2011.

USEPA. 2009c. Ag 101 Glossary. http://www.epa.gov/agriculture/ag101/glossary.html. (Verified January,
2011).

USEPA. 2010a. Annex 3: distribution and prevalence of selected zoonotic pathogens in U.S. domestic
livestock. *In* Quantitative microbial risk assessment to estimate illness in freshwater impacted by
agricultural animal sources of fecal contamination. EPA-822-R-10-005. USEPA, Office of Water,
Washington,
DC. http://water.epa.gov/scitech/swguidance/standards/criteria/health/recreation/upload/P4-
QMRA-508.pdf. (Verified December, 2011).

USEPA. 2010b. National lakes assessment: a collaborative survey of the nation's lakes. EPA-841-R-09-001.
USEPA, Office of Water, Washington,
DC. http://water.epa.gov/type/lakes/upload/nla_newlowres_fullrpt.pdf. (Verified December,
2011).

USEPA. 2011a. Inventory of U.S. greenhouse gas emissions and sinks: 1990-2009. EPA-430-R-11-005.
USEPA, Washington, DC. http://epa.gov/climatechange/emissions/usinventoryreport.html.
(Verified December, 2011).

USEPA. 2011b. Anaerobic digestion. AgStar. Updated November 9,
2011. http://www.epa.gov/agstar/anaerobic/index.html. (Verified December, 2011).

USEPA. 2011c. U.S. farm anaerobic digestion systems: a 2010
snapshot. http://www.epa.gov/agstar/documents/2010_digester_update.pdf. (Verified December,
2011).

United States Food and Drug Administration (USFDA). 1994. Environmental Assessment Report for
Zeranol. http://www.fda.gov/downloads/AnimalVeterinary/DevelopmentApprovalProcess/Enviro
nmentalAssessments/UCM071898.pdf. (Verified February, 2012).

USFDA. 1996. Environmental Assessment Report for NADA 34-
254. http://www.fda.gov/downloads/AnimalVeterinary/DevelopmentApprovalProcess/Environme
ntalAssessments/UCM071903.pdf. (Verified February, 2012).

USFDA. 2002. The use of steroid hormones for growth promotion in food-producing animals. FDA
Veterinarian Newsletter, Volume XVI, No
V. http://www.fda.gov/AnimalVeterinary/NewsEvents/FDAVeterinarianNewsletter/ucm110712.h
tm. (Verified December, 2011).

USFDA. 2009. Drugs@FDA. Accessed August,
2009. http://www.accessdata.fda.gov/scripts/cder/drugsatfda/. (Verified December, 2011).

USFDA. 2010. Summary report on antimicrobials sold or distributed for use in food-producing
animals. http://www.fda.gov/downloads/ForIndustry/UserFees/AnimalDrugUserFeeActADUFA/
UCM231851.pdf. (Verified December, 2011).

USFDA. 2011a. Summary report on antimicrobials sold or distributed for use in food-producing
animals. http://www.fda.gov/downloads/ForIndustry/UserFees/AnimalDrugUserFeeActADUFA/
UCM277657.pdf. (Verified February, 2012).

USFDA. 2011b. Animal Drugs@FDA. Accessed December,
2011. http://www.accessdata.fda.gov/scripts/animaldrugsatfda/. (Verified December, 2011).

USFDA. 2011c. Questions and answers regarding 3-Nitro (Roxarsone). Updated June 8,
2011. http://www.fda.gov/AnimalVeterinary/SafetyHealth/ProductSafetyInformation/ucm258313.
htm. (Verified December, 2011).

USFDA. 2011d. Steroid hormone implants used for growth in food-producing animals. Updated February 8,
2011. http://www.fda.gov/AnimalVeterinary/SafetyHealth/ProductSafetyInformation/ucm055436.
htm. (Verified December, 2011).

USFDA. 2011e. Bovine somatotropin (BST). Updated February 8,
2011. http://www.fda.gov/AnimalVeterinary/SafetyHealth/ProductSafetyInformation/ucm055435.
htm. (Verified December, 2011).

USFDA. 2012a. The Bad Bug Book: Foodborne Pathogenic Microorganisms and Natural Toxins
Handbook. http://www.fda.gov/downloads/Food/FoodSafety/FoodborneIllness/FoodborneIllnes
sFoodbornePathogensNaturalToxins/BadBugBook/UCM297627.pdf. (Verified November, 2012).

USFDA. 2012b. Cephalosporin order of prohibition questions and answers. Updated January 6,
2012. http://www.fda.gov/AnimalVeterinary/NewsEvents/CVMUpdates/ucm054434.htm.
(Verified February, 2012).

United States Government Accountability Office (USGAO). 1999. Food safety: the agricultural use of
antibiotics and its implications for human health. GAO-RCED-99-74. Report to the Committee on
Agriculture, Nutrition and Forestry, U.S. Senate. http://www.gao.gov/archive/1999/rc99074.pdf.
(Verified December, 2011).

USGAO. 2011a. Antibiotic resistance- agencies have made limited progress addressing antibiotic use in animals. GAO-11-801. Report to the Ranking Member, Committee on Rules, House of Representatives. http://www.gao.gov/new.items/d11801.pdf. (Verified December, 2011).

USGAO. 2011b. Environmental Protection Agency Major Management Challenges. Testimony before the Subcommittee on Interior, Environment, and Related Agencies, Committee on Appropriations, U.S. House of Representatives. http://www.gao.gov/assets/130/125556.pdf. (Verified July, 2012).

Valcour, J.E., P. Michel, S.A. McEwen, and J.B. Wilson. 2002. Associations between indicators of livestock farming intensity and incidence of human Shiga toxin-producing *Escherichia coli* infection. Emerging Infectious Diseases. 8(3):252-257.

van den Bogaard, A.E., R. Willems, N. London, J. Top, and E. Stobberingh. 2002. Antibiotic resistance of faecal enterococci in poultry, poultry farmers, and poultry slaughterers. Journal of Antimicrobial Chemotherapy. 49:497-505.

van Donsel, D.J., and E.E. Geldreich 1971. Relationship of *salmonellae* to fecal coliforms in bottom sediments. Water Research. 5: 1079-1087.

Varel, V.H., J.E. Wells, W.L. Shelver, C.P. Rice, D.L. Armstrong, and D.B. Parker. 2012. Effect of anaerobic digestion temperature on odour, coliforms and chlortetracycline in swine manure or monensin in cattle manure. Journal of Applied Microbiology. doi:10.1111/j.1365-2672.2012.05250.x.

Venglovsky J., N. Sasakova, and I. Placha. 2009. Pathogens and antibiotic residues in animal manures and hygienic and ecological risks related to subsequent land application. Bioresource Technology. 100(22): 5386-5391.

Vidon, P., M. Campbell, and M. Gray. 2008. Unrestricted cattle access to streams and water quality in till landscape of the Midwest. Agricultural Water Management. 95:322-330.

Vogt, R.L., H.E. Sours, T. Barrett, R.A. Feldman, R.J. Dickinson, and L. Witherell. 1982. Campylobacter enteritis associated with contaminated water. Annals of Internal Medicine. 96:292-296.

Wang, Q.H., M. Souza, J.A. Funk, W. Zhang, and L.J. Saiff. 2006. Prevalence of noroviruses and sapoviruses in swine of various ages determined by reverse transcription-PCR and microwell hybridization assays. Journal of Clinical Microbiology. 44(6):2057-2062.

Watanabe, N., B.A. Bergamaschi, K.A. Loftin, M.T. Meyer, and T. Harter. 2010. Use and environmental occurrence of antibiotics in freestall dairy farms with manured forage fields. Environmental Science & Technology. 44(17):6591-6600.

Weber A, J Potel, R Schäfer-Schmidt, A Prell, C Datzmann. 1995. Studies on the occurrence of Listeria monocytogenes in fecal samples of domestic and companion animals. Zentralbl Hyg Umweltmed. 198(2):117-23.

Weinberg, H.S., Z. Ye, and M.T. Meyer. 2004. Method development for the occurrence of residual antibiotics in drinking water. Water Resources Research Institute, University of North Carolina. UNC-WRRI-356. WRRI Project No. 50307.

Weinberg, H.S., V.J. Pereira, and Z. Ye. 2008. Drugs in drinking water: treatment options. p. 217-228. *In* D.S. Aga (ed.) Fate and transport of pharmaceuticals in the environment and water treatment systems. 1st ed. CRC Press, Boca Raton, FL.

Westerhoff, P., Y. Yoon, S. Snyder, and E. Wert. 2005. Fate of endocrine-disruptor, pharmaceutical, and personal care product chemicals during simulated drinking water treatment process. Environmental Science & Technology. 39(17):6649-6663.

Wilkes, G., T. Edge, V. Gannon, C. Jokinen, E. Lyautev, D. Medeiros, N. Neumann, N. Ruecker, E. Topp, and D.R. Lapen. 2009. Seasonal relationships among indicator bacteria, pathogenic bacteria, *Cryptosporidium* oocysts, *Giardia* cysts, and hydrological indices for surface waters within an agricultural landscape. Water Research. 43:2209-2223.

Withers, P.J.A., H. McDonald, K.A. Smith, and C.G. Chumbly. 1998. Behavior and impact of cow slurry beneath a storage lagoon: 1. groundwater contamination 1975-1982. Water Air and Soil Pollution. 107(1/4):35-49.

World Health Organization (WHO). 2000. Report on infectious diseases, overcoming antimicrobial resistance. http://www.who.int/infectious-disease-report/. (Verified December, 2011).

Xi, C., Y. Zhang, C. F. Marrs, W. Ye, C. Simon, B. Foxman, and J. Nriagu. 2009. Prevalence of antibiotic resistance in drinking water treatment and distribution systems. Applied and Environmental Microbiology. 75(17):5714-5718.

Yu, H., and J.G. Bruno. 1996. Immunomagnetic-electrochemiluminescent detection of *Escherichia coli* O157 and *Salmonella Typhimurium* in foods and environmental water samples. Applied and Environmental Microbiology. 62(2):587-592.

Zhao, Z., K.F. Knowlton, and N.G. Love. 2008. Hormones in waste from concentrated animal feeding operations. p. 291-329. *In* D.S. Aga (ed.) Fate and transport of pharmaceuticals in the environment and water treatment systems. 1st ed. CRC Press, Boca Raton, FL.

Zheng, W., S.R. Yates, and S.A. Bradford. 2008. Analysis of steroid hormones in a typical dairy waste disposal system. Environmental Science & Technology. 42(2):530-535.

Ziemer,C., J. Bonner, D. Cole, J. Vinjé, V. Constantini, S. Goyal, M. Gramer, R. Mackie, X. Meng, G. Myers, and L. Saif, L. 2010. Fate and transport of zoonotic, bacterial, viral, and parasitic pathogens during swine manure treatment, storage, and land application. Journal of Animal Science. 88:E84-E94.

Zounková, R., Z. Klimešová, L. Nepejchalová, K. Hilscherová, and L. Bláha. 2011. Complex evaluation of ecotoxicity and genotoxicity of antimicrobials oxytetracycline and flumequine used in aquaculture. Environmental Toxicology and Chemistry. 30(5):1184-1189.

# Appendix 1. Livestock Animal Unit and Manure Production Calculations

Livestock manure production was estimated using standard methods and conversion factors developed by the USDA's NRCS (see for example Kellogg et al. 2000, Gollehon et al. 2001, and Midwest Planning Service 1985), converting livestock and poultry head counts to animal units (AU). Animal units are a common unit of measure based on animal weight, allowing for the calculation of manure generation and a method for aggregating across animal types and life stages. For this report we used USDA's 2007 Census of Agriculture livestock count data for cattle, swine, chickens (layers and broilers), and turkeys as well as acreage of land in farms for each state. "Land in farms" is defined by the USDA (2009a) as primarily agricultural land used for grazing, pasture, or crops, but it may also include woodland and wasteland that is not under cultivation or used for grazing or pasture, provided it is on the farm operator's operation. For cattle, three categories were used: beef cows, milk cows, and "cattle excluding cows" (e.g., breeding and replacement stock). The total inventory numbers (head of animals) from the end of December, 2007 were used to generate the final numbers of AUs in each state. Similar, but more complex, methods were employed by Kellogg et al. (2000) which used USDA's Census of Agriculture data to calculate livestock and poultry manure generation and manure nutrient contributions, evaluate trends in livestock production, and quantify the extent to which manure nutrient contributions exceed crop assimilative capacity. Additionally, Kellogg et al. (2000) calculated AUs using 16 livestock categories/life stages from more detailed marketing statistics to refine estimates of manure generation and nutrient recovery, and make estimates of confinement operations. (Note: the overall state and national estimates of this report are within a few percentage points of the estimates of these reports for total manure generated).

The AU and manure production conversion factors were then related to the appropriate animals for breeding and marketing for each livestock type (see Table A-1). Following the procedures, three quarters of the "cattle excluding cows" were treated as "Steers, Calves, & Bulls" and the remaining quarter were treated as "Heifers & Dairy Calves," which assumes that roughly half of the animals in this category are adult animals slated for slaughter, and the remaining half is equally split between young females (heifers) and males (steers). Turkey counts were treated as slaughters to provide a more conservative estimate for this animal type (i.e., there are more AUs per slaughter turkey than breeder turkey, therefore providing lower manure generation estimates; see Table A-1).

**Table A-1. The number of animal units (AU) and associated manure generation per animal type as defined by USDA's NRCS.**

| Animal Type | Animals per AU | Manure Generation per AU (tons) |
|---|---|---|
| Beef Cattle | 1 | 11.5 |
| Dairy Cows | 0.74 | 15.24 |
| Heifers & Dairy Calves | 1.82 | 12.05 |
| Steers, Calves, & Bulls | 1.64 | 10.59 |
| Swine, Breeders | 2.67 | 6.11 |
| Swine, Market | 9.09 | 14.69 |
| Chickens, Layers | 250 | 11.45 |
| Chickens, Broilers | 455 | 14.97 |
| Turkeys for Slaughter | 67 | 8.18 |
| Turkeys Hens for Breeding | 50 | 8.18 |

*Kellogg et al. 2000, Gollehon et al. 2001.*

Converting all the animal types to AUs allows the total number of all AUs to be summed as well as the total estimated manure produced to be summed, so a "total" comparison among the states can be done, as shown in the tables in this appendix. Also, livestock and poultry manure generation was estimated by dividing state manure generation by the sum of land in farms both owned and rented in each state – the most likely land-base for the application of the manure – using data from the USDA's 2007 Census of Agriculture, as discussed in Chapter 2. To illustrate the AU and manure generation calculations, the following example is provided using beef cattle counts in Texas. Calculated data for all states are shown in Tables A-2 to A-9.

$$\text{AUs} = \frac{2007 \; USDA \; Census \; of \; Agriculture \; inventory}{Animals \; per \; AU}$$

$$Texas \; beef \; AUs = \frac{5,259,843}{1} = 5,259,843 \; AUs$$

$$Percentage \; of \; U.S. \; livestock = \frac{State \; inventory}{Sum \; of \; inventory \; in \; all \; reporting \; states} \times 100$$

$$Texas' percentage \; of \; U.S. \; beef \; stock = \frac{5,259,843}{32,834,801} \times 100 = 16.02\%$$

$$Tons \; of \; manure \; produced = AUs \times tons \; manure \; produced \; per \; AU$$

$$Texas' beef \; manure \; production = 5,259,843 \times 11.5 = 60,488,195 \; tons$$

$$Total \; AUs = (Beef \; Cow + Milk \; Cow + Cattle \; Excluding \; Cows \; AUs)$$
$$+ (Swine \; Breeder + Swine \; Market \; AUs) + (Layer + Broiler \; AUs) + (Turkey \; AUs)$$

$$Texas' AUs = (5,259,843 + 404,399 + 4,784,377) + (35,550 + 116,708) + (76,467 + 260,686)$$
$$+ (29,654) = 11,109,770 \; AUs$$

$$\frac{AUs \; or \; tons \; manure}{acre} = \frac{AUs \; or \; tons \; manure}{land \; owned, in \; farms + land, rented, in \; farms}$$

$$\frac{AUs}{acre} \; in \; Texas = \frac{11,109,770}{75,578,240 + 54,299,426} = 0.09 \frac{AUs}{acre}$$

Tables A-2 through A-9 present summaries of livestock AUs and estimated total manure generated by those livestock for all 50 states. The states are listed in rank-order in the different categories.

**Table A-2. Total animal units and estimated tons of manure produced for beef and dairy cattle in 2007.**

| National Rank | State | Total Beef Cattle AUs | Percent of Total Beef Cattle AUs | Total Tons Manure | National Rank | State | Total Dairy Cow AUs | Percent of Total Dairy Cow AUs | Total Tons Manure |
|---|---|---|---|---|---|---|---|---|---|
| 1 | TEXAS | 5,259,843 | 16.02% | 60,488,195 | 1 | CALIFORNIA | 2,487,473 | 19.86% | 37,909,088 |
| 2 | MISSOURI | 2,089,181 | 6.36% | 24,025,582 | 2 | WISCONSIN | 1,688,255 | 13.48% | 25,729,012 |
| 3 | OKLAHOMA | 2,063,613 | 6.28% | 23,731,550 | 3 | NEW YORK | 846,561 | 6.76% | 12,901,587 |
| 4 | NEBRASKA | 1,889,842 | 5.76% | 21,733,183 | 4 | PENNSYLVANIA | 747,731 | 5.97% | 11,395,422 |
| 5 | SOUTH DAKOTA | 1,649,492 | 5.02% | 18,969,158 | 5 | IDAHO | 724,950 | 5.79% | 11,048,238 |
| 6 | MONTANA | 1,522,187 | 4.64% | 17,505,151 | 6 | MINNESOTA | 621,286 | 4.96% | 9,468,406 |
| 7 | KANSAS | 1,516,374 | 4.62% | 17,438,301 | 7 | TEXAS | 546,485 | 4.36% | 8,328,433 |
| 8 | TENNESSEE | 1,179,102 | 3.59% | 13,559,673 | 8 | MICHIGAN | 465,180 | 3.71% | 7,089,339 |
| 9 | KENTUCKY | 1,166,385 | 3.55% | 13,413,428 | 9 | NEW MEXICO | 441,081 | 3.52% | 6,722,076 |
| 10 | ARKANSAS | 947,765 | 2.89% | 10,899,298 | 10 | OHIO | 367,484 | 2.93% | 5,600,453 |
| 11 | FLORIDA | 942,419 | 2.87% | 10,837,819 | 11 | WASHINGTON | 328,557 | 2.62% | 5,007,205 |
| 12 | NORTH DAKOTA | 930,023 | 2.83% | 10,695,265 | 12 | IOWA | 291,069 | 2.32% | 4,435,890 |
| 13 | IOWA | 904,100 | 2.75% | 10,397,150 | 13 | ARIZONA | 248,303 | 1.98% | 3,784,133 |
| 14 | COLORADO | 735,014 | 2.24% | 8,452,661 | 14 | INDIANA | 224,526 | 1.79% | 3,421,771 |
| 15 | WYOMING | 732,141 | 2.23% | 8,419,622 | 15 | VERMONT | 188,809 | 1.51% | 2,877,456 |
| 16 | VIRGINIA | 695,061 | 2.12% | 7,993,202 | 16 | COLORADO | 171,546 | 1.37% | 2,614,360 |
| 17 | ALABAMA | 678,949 | 2.07% | 7,807,914 | 17 | FLORIDA | 161,968 | 1.29% | 2,468,386 |
| 18 | CALIFORNIA | 662,423 | 2.02% | 7,617,865 | 18 | OREGON | 157,822 | 1.26% | 2,405,202 |
| 19 | OREGON | 604,069 | 1.84% | 6,946,794 | 19 | KANSAS | 156,262 | 1.25% | 2,381,435 |
| 20 | GEORGIA | 554,099 | 1.69% | 6,372,139 | 20 | MISSOURI | 149,132 | 1.19% | 2,272,778 |
| 21 | NEW MEXICO | 530,173 | 1.61% | 6,096,990 | 21 | ILLINOIS | 134,699 | 1.08% | 2,052,807 |
| 22 | MISSISSIPPI | 521,517 | 1.59% | 5,997,446 | 22 | VIRGINIA | 133,672 | 1.07% | 2,037,156 |
| 23 | LOUISIANA | 510,837 | 1.56% | 5,874,626 | 23 | KENTUCKY | 122,246 | 0.98% | 1,863,028 |
| 24 | IDAHO | 476,292 | 1.45% | 5,477,358 | 24 | SOUTH DAKOTA | 116,545 | 0.93% | 1,776,140 |
| 25 | ILLINOIS | 429,111 | 1.31% | 4,934,777 | 25 | UTAH | 115,219 | 0.92% | 1,755,936 |
| 26 | MINNESOTA | 399,768 | 1.22% | 4,597,332 | 26 | GEORGIA | 104,315 | 0.83% | 1,589,759 |
| 27 | NORTH CAROLINA | 373,024 | 1.14% | 4,289,776 | 27 | OKLAHOMA | 89,220 | 0.71% | 1,359,717 |
| 28 | UTAH | 364,744 | 1.11% | 4,194,556 | 28 | TENNESSEE | 82,609 | 0.66% | 1,258,968 |
| 29 | OHIO | 293,757 | 0.89% | 3,378,206 | 29 | MARYLAND | 77,259 | 0.62% | 1,177,434 |
| 30 | WASHINGTON | 274,001 | 0.83% | 3,151,012 | 30 | NEBRASKA | 73,527 | 0.59% | 1,120,552 |
| 31 | WISCONSIN | 269,820 | 0.82% | 3,102,930 | 31 | NORTH CAROLINA | 64,309 | 0.51% | 980,076 |
| 32 | NEVADA | 238,662 | 0.73% | 2,744,613 | 32 | MAINE | 43,955 | 0.35% | 669,880 |
| 33 | INDIANA | 235,299 | 0.72% | 2,705,939 | 33 | LOUISIANA | 38,295 | 0.31% | 583,610 |
| 34 | SOUTH CAROLINA | 230,419 | 0.70% | 7,649,819 | 34 | NEVADA | 37,378 | 0.30% | 569,646 |
| 35 | WEST VIRGINIA | 203,711 | 0.62% | 2,342,677 | 35 | NORTH DAKOTA | 35,782 | 0.29% | 545,324 |
| 36 | ARIZONA | 197,060 | 0.60% | 2,266,190 | 36 | MISSISSIPPI | 30,486 | 0.24% | 464,614 |
| 37 | PENNSYLVANIA | 158,430 | 0.48% | 1,821,945 | 37 | CONNECTICUT | 27,953 | 0.22% | 425,999 |
| 38 | MICHIGAN | 109,500 | 0.33% | 1,259,250 | 38 | SOUTH CAROLINA | 24,095 | 0.19% | 367,202 |
| 39 | NEW YORK | 103,620 | 0.32% | 1,191,630 | 39 | MONTANA | 23,592 | 0.19% | 359,540 |
| 40 | HAWAII | 86,000 | 0.26% | 989,000 | 40 | ARKANSAS | 22,592 | 0.18% | 344,300 |
| 41 | MARYLAND | 44,015 | 0.13% | 506,173 | 41 | MASSACHUSETTS | 20,338 | 0.16% | 309,949 |
| 42 | MAINE | 12,114 | 0.04% | 139,311 | 42 | NEW HAMPSHIRE | 19,745 | 0.16% | 300,908 |
| 43 | VERMONT | 10,002 | 0.03% | 115,023 | 43 | ALABAMA | 17,516 | 0.14% | 266,947 |
| 44 | NEW JERSEY | 9,298 | 0.03% | 106,927 | 44 | WEST VIRGINIA | 15,870 | 0.13% | 241,863 |
| 45 | MASSACHUSETTS | 8,646 | 0.03% | 99,429 | 45 | NEW JERSEY | 13,230 | 0.11% | 201,621 |
| 46 | ALASKA | 6,468 | 0.02% | 74,382 | 46 | WYOMING | 8,978 | 0.07% | 136,830 |
| 47 | CONNECTICUT | 5,982 | 0.02% | 68,793 | 47 | DELAWARE | 8,819 | 0.07% | 134,400 |
| 48 | NEW HAMPSHIRE | 4,981 | 0.02% | 57,282 | 48 | HAWAII | 3,103 | 0.02% | 47,285 |
| 49 | DELAWARE | 3,668 | 0.01% | 42,182 | 49 | RHODE ISLAND | 1,791 | 0.01% | 27,288 |
| 50 | RHODE ISLAND | 1,800 | 0.01% | 20,700 | 50 | ALASKA | 780 | 0.01% | 11,883 |
| | **U.S. TOTAL** | **32,834,801** | | **377,600,212** | | **U.S. TOTAL** | **12,522,397** | | **190,841,335** |

*USDA 2009a.*

**Table A-3. Total animal units and estimated tons of manure produced for cattle other than beef and dairy cattle and for all cattle combined in 2007.**

| National Rank | State | Other Cattle AUs | Percent of Total Other Cattle AUs | Total Tons Manure | National Rank | State | Total Cattle AUs | Percent of Total Cattle AUs | Total Tons Manure |
|---|---|---|---|---|---|---|---|---|---|
| 1 | TEXAS | 4,784,377 | 14.83% | 52,280,036 | 1 | TEXAS | 10,590,705 | 13.64% | 121,096,664 |
| 2 | KANSAS | 2,995,494 | 9.29% | 32,732,479 | 2 | CALIFORNIA | 4,930,886 | 6.35% | 64,988,253 |
| 3 | NEBRASKA | 2,754,972 | 8.54% | 30,104,233 | 3 | NEBRASKA | 4,718,341 | 6.08% | 52,957,968 |
| 4 | OKLAHOMA | 1,939,667 | 6.01% | 21,195,210 | 4 | KANSAS | 4,668,130 | 6.01% | 52,552,215 |
| 5 | CALIFORNIA | 1,780,990 | 5.52% | 19,461,300 | 5 | OKLAHOMA | 4,092,501 | 5.27% | 46,286,477 |
| 6 | IOWA | 1,702,481 | 5.28% | 18,603,413 | 6 | MISSOURI | 3,483,075 | 4.49% | 39,900,167 |
| 7 | MISSOURI | 1,244,762 | 3.86% | 13,601,808 | 7 | WISCONSIN | 3,061,084 | 3.94% | 40,884,779 |
| 8 | SOUTH DAKOTA | 1,160,811 | 3.60% | 12,684,456 | 8 | SOUTH DAKOTA | 2,926,847 | 3.77% | 33,429,753 |
| 9 | COLORADO | 1,119,957 | 3.47% | 12,238,042 | 9 | IOWA | 2,897,650 | 3.73% | 33,436,453 |
| 10 | WISCONSIN | 1,103,008 | 3.42% | 12,052,836 | 10 | MONTANA | 2,170,213 | 2.80% | 24,688,030 |
| 11 | MINNESOTA | 913,248 | 2.83% | 9,979,278 | 11 | COLORADO | 2,026,517 | 2.61% | 23,305,063 |
| 12 | IDAHO | 727,526 | 2.26% | 7,949,855 | 12 | KENTUCKY | 1,965,738 | 2.53% | 22,675,367 |
| 13 | KENTUCKY | 677,107 | 2.10% | 7,398,911 | 13 | MINNESOTA | 1,934,302 | 2.49% | 24,045,016 |
| 14 | MONTANA | 624,434 | 1.94% | 6,823,339 | 14 | IDAHO | 1,928,768 | 2.48% | 24,475,451 |
| 15 | PENNSYLVANIA | 533,663 | 1.65% | 5,831,465 | 15 | TENNESSEE | 1,786,091 | 2.30% | 20,548,663 |
| 16 | TENNESSEE | 524,380 | 1.63% | 5,730,022 | 16 | FLORIDA | 1,490,177 | 1.92% | 17,521,825 |
| 17 | NORTH DAKOTA | 508,464 | 1.58% | 5,556,104 | 17 | NORTH DAKOTA | 1,474,269 | 1.90% | 16,796,693 |
| 18 | ARKANSAS | 498,443 | 1.55% | 5,446,603 | 18 | ARKANSAS | 1,468,800 | 1.89% | 16,690,201 |
| 19 | VIRGINIA | 459,235 | 1.42% | 5,018,169 | 19 | PENNSYLVANIA | 1,439,824 | 1.86% | 19,048,832 |
| 20 | NEW YORK | 424,139 | 1.31% | 4,634,665 | 20 | NEW YORK | 1,374,319 | 1.77% | 18,727,881 |
| 21 | OHIO | 420,264 | 1.30% | 4,592,329 | 21 | NEW MEXICO | 1,369,334 | 1.76% | 17,168,985 |
| 22 | ILLINOIS | 417,654 | 1.29% | 4,563,802 | 22 | VIRGINIA | 1,287,967 | 1.66% | 15,048,526 |
| 23 | NEW MEXICO | 398,080 | 1.23% | 4,349,920 | 23 | OREGON | 1,159,334 | 1.49% | 13,694,955 |
| 24 | OREGON | 397,443 | 1.23% | 4,342,960 | 24 | WYOMING | 1,081,879 | 1.39% | 12,280,016 |
| 25 | FLORIDA | 385,790 | 1.20% | 4,215,621 | 25 | OHIO | 1,081,505 | 1.39% | 13,570,987 |
| 26 | ARIZONA | 368,246 | 1.14% | 4,023,911 | 26 | ALABAMA | 990,986 | 1.28% | 11,293,163 |
| 27 | MICHIGAN | 353,521 | 1.10% | 3,863,009 | 27 | ILLINOIS | 981,463 | 1.26% | 11,551,386 |
| 28 | WYOMING | 340,760 | 1.06% | 3,723,564 | 28 | GEORGIA | 947,306 | 1.22% | 11,118,694 |
| 29 | WASHINGTON | 339,986 | 1.05% | 3,715,110 | 29 | WASHINGTON | 942,544 | 1.21% | 11,873,326 |
| 30 | ALABAMA | 294,521 | 0.91% | 3,218,302 | 30 | MICHIGAN | 928,201 | 1.20% | 12,211,598 |
| 31 | GEORGIA | 288,892 | 0.90% | 3,156,797 | 31 | MISSISSIPPI | 815,604 | 1.05% | 9,342,488 |
| 32 | INDIANA | 281,820 | 0.87% | 3,079,514 | 32 | ARIZONA | 813,609 | 1.05% | 10,074,234 |
| 33 | MISSISSIPPI | 263,601 | 0.82% | 2,880,428 | 33 | LOUISIANA | 751,019 | 0.97% | 8,664,305 |
| 34 | NORTH CAROLINA | 237,616 | 0.74% | 2,596,482 | 34 | INDIANA | 741,645 | 0.96% | 9,207,223 |
| 35 | UTAH | 233,987 | 0.73% | 2,556,837 | 35 | UTAH | 713,950 | 0.92% | 8,507,329 |
| 36 | LOUISIANA | 201,887 | 0.63% | 2,206,070 | 36 | NORTH CAROLINA | 674,949 | 0.87% | 7,866,334 |
| 37 | WEST VIRGINIA | 116,303 | 0.36% | 1,270,874 | 37 | NEVADA | 380,292 | 0.49% | 4,453,441 |
| 38 | NEVADA | 104,252 | 0.32% | 1,139,181 | 38 | SOUTH CAROLINA | 345,349 | 0.44% | 4,009,602 |
| 39 | SOUTH CAROLINA | 90,836 | 0.28% | 992,582 | 39 | WEST VIRGINIA | 335,885 | 0.43% | 3,855,413 |
| 40 | VERMONT | 68,449 | 0.21% | 747,957 | 40 | VERMONT | 267,260 | 0.34% | 3,740,436 |
| 41 | MARYLAND | 53,115 | 0.16% | 580,400 | 41 | MARYLAND | 174,389 | 0.22% | 2,264,007 |
| 42 | HAWAII | 37,574 | 0.12% | 410,576 | 42 | HAWAII | 126,676 | 0.16% | 1,446,861 |
| 43 | MAINE | 25,898 | 0.08% | 282,997 | 43 | MAINE | 81,968 | 0.11% | 1,092,188 |
| 44 | CONNECTICUT | 14,002 | 0.04% | 153,007 | 44 | CONNECTICUT | 47,937 | 0.06% | 647,799 |
| 45 | MASSACHUSETTS | 13,770 | 0.04% | 150,472 | 45 | MASSACHUSETTS | 42,754 | 0.06% | 559,850 |
| 46 | NEW JERSEY | 11,364 | 0.04% | 124,181 | 46 | NEW HAMPSHIRE | 35,006 | 0.05% | 470,530 |
| 47 | NEW HAMPSHIRE | 10,281 | 0.03% | 112,341 | 47 | NEW JERSEY | 33,892 | 0.04% | 432,729 |
| 48 | DELAWARE | 6,423 | 0.02% | 70,181 | 48 | DELAWARE | 18,909 | 0.02% | 246,763 |
| 49 | ALASKA | 4,625 | 0.01% | 50,543 | 49 | ALASKA | 11,873 | 0.02% | 136,808 |
| 50 | RHODE ISLAND | 1,166 | 0.00% | 12,736 | 50 | RHODE ISLAND | 4,756 | 0.01% | 60,724 |
| | U.S. TOTAL | 32,259,283 | | 352,504,907 | | U.S. TOTAL | 77,616,481 | | 920,946,454 |

*USDA 2009a.*

## Table A-4. Total animal units and estimated tons of manure produced for breeder and market hogs in 2007.

| National Rank | State | Total Breeder Hog AUs | Percent of Total Breeder Hog AUs | Total Tons Manure | National Rank | State | Total Market Hog AUs | Percent of Total Market Hog AUs | Total Tons Manure |
|---|---|---|---|---|---|---|---|---|---|
| 1 | IOWA | 406,815 | 17.77% | 2,485,637 | 1 | IOWA | 2,003,179 | 30.25% | 29,426,699 |
| 2 | NORTH CAROLINA | 378,608 | 16.54% | 2,313,294 | 2 | NORTH CAROLINA | 1,003,644 | 15.16% | 14,743,526 |
| 3 | MINNESOTA | 223,606 | 9.77% | 1,366,235 | 3 | MINNESOTA | 776,156 | 11.72% | 11,401,727 |
| 4 | ILLINOIS | 191,057 | 8.34% | 1,167,360 | 4 | ILLINOIS | 416,787 | 6.29% | 6,122,600 |
| 5 | OKLAHOMA | 147,216 | 6.43% | 899,488 | 5 | INDIANA | 369,134 | 5.57% | 5,422,574 |
| 6 | NEBRASKA | 145,798 | 6.37% | 890,824 | 6 | NEBRASKA | 316,751 | 4.78% | 4,653,068 |
| 7 | MISSOURI | 134,134 | 5.86% | 819,557 | 7 | MISSOURI | 301,797 | 4.56% | 4,433,394 |
| 8 | INDIANA | 117,465 | 5.13% | 717,712 | 8 | OKLAHOMA | 220,606 | 3.33% | 3,240,698 |
| 9 | KANSAS | 69,309 | 3.03% | 423,476 | 9 | KANSAS | 187,040 | 2.82% | 2,747,625 |
| 10 | COLORADO | 62,551 | 2.73% | 382,189 | 10 | OHIO | 183,864 | 2.78% | 2,700,956 |
| 11 | SOUTH DAKOTA | 61,980 | 2.71% | 378,699 | 11 | SOUTH DAKOTA | 145,715 | 2.20% | 2,140,550 |
| 12 | OHIO | 59,837 | 2.61% | 365,602 | 12 | TEXAS | 116,708 | 1.76% | 1,714,435 |
| 13 | PENNSYLVANIA | 44,924 | 1.96% | 274,483 | 13 | PENNSYLVANIA | 115,237 | 1.74% | 1,692,829 |
| 14 | MICHIGAN | 39,404 | 1.72% | 240,759 | 14 | MICHIGAN | 101,963 | 1.54% | 1,497,839 |
| 15 | TEXAS | 35,550 | 1.55% | 217,209 | 15 | COLORADO | 78,733 | 1.19% | 1,156,588 |
| 16 | ARKANSAS | 31,477 | 1.37% | 192,325 | 16 | WISCONSIN | 42,260 | 0.64% | 620,802 |
| 17 | WISCONSIN | 19,726 | 0.86% | 120,527 | 17 | VIRGINIA | 37,293 | 0.56% | 547,827 |
| 18 | GEORGIA | 16,521 | 0.72% | 100,943 | 18 | KENTUCKY | 33,627 | 0.51% | 493,980 |
| 19 | KENTUCKY | 15,863 | 0.69% | 96,922 | 19 | SOUTH CAROLINA | 29,266 | 0.44% | 429,918 |
| 20 | NORTH DAKOTA | 14,302 | 0.62% | 87,384 | 20 | GEORGIA | 24,132 | 0.36% | 354,499 |
| 21 | VIRGINIA | 12,055 | 0.53% | 73,656 | 21 | ARKANSAS | 22,585 | 0.34% | 331,774 |
| 22 | WYOMING | 10,416 | 0.45% | 63,642 | 22 | ALABAMA | 17,600 | 0.27% | 258,544 |
| 23 | SOUTH CAROLINA | 10,399 | 0.45% | 63,537 | 23 | NORTH DAKOTA | 15,786 | 0.24% | 231,894 |
| 24 | CALIFORNIA | 8,001 | 0.35% | 48,889 | 24 | CALIFORNIA | 14,590 | 0.22% | 214,320 |
| 25 | ALABAMA | 6,851 | 0.30% | 41,857 | 25 | TENNESSEE | 13,778 | 0.21% | 202,396 |
| 26 | NEW YORK | 5,005 | 0.22% | 30,580 | 26 | WYOMING | 8,731 | 0.13% | 128,265 |
| 27 | TENNESSEE | 4,857 | 0.21% | 29,674 | 27 | NEW YORK | 7,962 | 0.12% | 116,967 |
| 28 | IDAHO | 2,282 | 0.10% | 13,941 | 28 | IDAHO | 2,938 | 0.04% | 43,152 |
| 29 | FLORIDA | 2,025 | 0.09% | 12,371 | 29 | WASHINGTON | 2,643 | 0.04% | 38,823 |
| 30 | WASHINGTON | 1,694 | 0.07% | 10,348 | 30 | OREGON | 1,891 | 0.03% | 27,778 |
| 31 | MARYLAND | 1,619 | 0.07% | 9,895 | 31 | FLORIDA | 1,599 | 0.02% | 23,483 |
| 32 | OREGON | 1,474 | 0.06% | 9,007 | 32 | HAWAII | 1,217 | 0.02% | 17,870 |
| 33 | HAWAII | 1,451 | 0.06% | 8,868 | 33 | MASSACHUSETTS | 1,033 | 0.02% | 15,175 |
| 34 | DELAWARE | 961 | 0.04% | 5,870 | 34 | LOUISIANA | 911 | 0.01% | 13,379 |
| 35 | LOUISIANA | 875 | 0.04% | 5,346 | 35 | NEW JERSEY | 831 | 0.01% | 12,201 |
| 36 | MASSACHUSETTS | 810 | 0.04% | 4,950 | 36 | WEST VIRGINIA | 814 | 0.01% | 11,959 |
| 37 | WEST VIRGINIA | 580 | 0.03% | 3,542 | 37 | DELAWARE | 703 | 0.01% | 10,327 |
| 38 | NEW JERSEY | 375 | 0.02% | 2,291 | 38 | MAINE | 381 | 0.01% | 5,592 |
| 39 | CONNECTICUT | 354 | 0.02% | 2,160 | 39 | CONNECTICUT | 297 | 0.00% | 4,365 |
| 40 | MAINE | 352 | 0.02% | 2,153 | 40 | NEW HAMPSHIRE | 242 | 0.00% | 3,557 |
| 41 | NEVADA | 284 | 0.01% | 1,735 | 41 | NEVADA | 241 | 0.00% | 3,541 |
| 42 | NEW HAMPSHIRE | 221 | 0.01% | 1,352 | 42 | VERMONT | 240 | 0.00% | 3,533 |
| 43 | NEW MEXICO | 219 | 0.01% | 1,339 | 43 | RHODE ISLAND | 196 | 0.00% | 2,881 |
| 44 | RHODE ISLAND | 200 | 0.01% | 1,220 | 44 | NEW MEXICO | 153 | 0.00% | 2,241 |
| 45 | VERMONT | 193 | 0.01% | 1,179 | 45 | ALASKA | D | | |
| 46 | ALASKA | D | | | 46 | ARIZONA | D | | |
| 47 | ARIZONA | D | | | 47 | MARYLAND | D | | |
| 48 | MISSISSIPPI | D | | | 48 | MISSISSIPPI | D | | |
| 49 | MONTANA | D | | | 49 | MONTANA | D | | |
| 50 | UTAH | D | | | 50 | UTAH | D | | |
| | U.S. TOTAL | 2,289,694 | | 13,990,028 | | U.S. TOTAL | 6,621,249 | | 97,266,149 |

*"D" signifies that the data were not disclosed (because there were too few producers in the category to protect confidentiality).*
*USDA 2009a.*

**Table A-5. Total animal units and estimated tons of manure produced for swine (breeder and market hogs combined) in 2007.**

| National Rank | State | Total Swine AUs | Percent of Total Swine AUs | Total Tons Manure |
|---|---|---|---|---|
| 1 | IOWA | 2,409,994 | 27.05% | 31,912,337 |
| 2 | NORTH CAROLINA | 1,382,252 | 15.51% | 17,056,820 |
| 3 | MINNESOTA | 999,762 | 11.22% | 12,767,962 |
| 4 | ILLINOIS | 607,844 | 6.82% | 7,289,960 |
| 5 | INDIANA | 486,599 | 5.46% | 6,140,286 |
| 6 | NEBRASKA | 462,548 | 5.19% | 5,543,892 |
| 7 | MISSOURI | 435,930 | 4.89% | 5,252,950 |
| 8 | OKLAHOMA | 367,821 | 4.13% | 4,140,186 |
| 9 | KANSAS | 256,349 | 2.88% | 3,171,100 |
| 10 | OHIO | 243,700 | 2.73% | 3,066,558 |
| 11 | SOUTH DAKOTA | 207,695 | 2.33% | 2,519,248 |
| 12 | PENNSYLVANIA | 160,160 | 1.80% | 1,967,313 |
| 13 | TEXAS | 152,257 | 1.71% | 1,931,644 |
| 14 | MICHIGAN | 141,367 | 1.59% | 1,738,598 |
| 15 | COLORADO | 141,284 | 1.59% | 1,538,776 |
| 16 | WISCONSIN | 61,986 | 0.70% | 741,329 |
| 17 | ARKANSAS | 54,062 | 0.61% | 524,100 |
| 18 | KENTUCKY | 49,490 | 0.56% | 590,902 |
| 19 | VIRGINIA | 49,348 | 0.55% | 621,484 |
| 20 | GEORGIA | 40,653 | 0.46% | 455,442 |
| 21 | SOUTH CAROLINA | 39,665 | 0.45% | 493,455 |
| 22 | NORTH DAKOTA | 30,088 | 0.34% | 319,278 |
| 23 | ALABAMA | 24,451 | 0.27% | 300,401 |
| 24 | CALIFORNIA | 22,591 | 0.25% | 263,210 |
| 25 | WYOMING | 19,148 | 0.21% | 191,908 |
| 26 | TENNESSEE | 18,634 | 0.21% | 232,069 |
| 27 | NEW YORK | 12,967 | 0.15% | 147,547 |
| 28 | IDAHO | 5,219 | 0.06% | 57,093 |
| 29 | WASHINGTON | 4,336 | 0.05% | 49,171 |
| 30 | FLORIDA | 3,623 | 0.04% | 35,854 |
| 31 | OREGON | 3,365 | 0.04% | 36,786 |
| 32 | HAWAII | 2,668 | 0.03% | 26,738 |
| 33 | MASSACHUSETTS | 1,843 | 0.02% | 20,125 |
| 34 | LOUISIANA | 1,786 | 0.02% | 18,725 |
| 35 | DELAWARE | 1,664 | 0.02% | 16,196 |
| 36 | MARYLAND | 1,619 | 0.02% | 9,895 |
| 37 | WEST VIRGINIA | 1,394 | 0.02% | 15,501 |
| 38 | NEW JERSEY | 1,205 | 0.01% | 14,492 |
| 39 | MAINE | 733 | 0.01% | 7,745 |
| 40 | CONNECTICUT | 651 | 0.01% | 6,525 |
| 41 | NEVADA | 525 | 0.01% | 5,275 |
| 42 | NEW HAMPSHIRE | 463 | 0.01% | 4,909 |
| 43 | VERMONT | 433 | 0.00% | 4,711 |
| 44 | RHODE ISLAND | 396 | 0.00% | 4,101 |
| 45 | NEW MEXICO | 372 | 0.00% | 3,580 |
| 46 | ALASKA | D | | |
| 47 | ARIZONA | D | | |
| 48 | MISSISSIPPI | D | | |
| 49 | MONTANA | D | | |
| 50 | UTAH | D | | |
| | U.S. TOTAL | 8,910,943 | | 111,256,177 |

*"D" signifies that the data were not disclosed (because there were too few producers in the category to protect confidentiality).*
*USDA 2009a.*

**Table A-6. Total animal units and estimated tons of manure produced for broiler and layer chickens in 2007.**

| National Rank | State | Total Broiler AUs | Percent of Total Broiler AUs | Total Tons Manure | National Rank | State | Total Layer AUs | Percent of Total Layer AUs | Total Tons Manure |
|---|---|---|---|---|---|---|---|---|---|
| 1 | GEORGIA | 517,363 | 14.69% | 7,744,926 | 1 | IOWA | 215,175 | 15.92% | 2,463,752 |
| 2 | ARKANSAS | 444,830 | 12.63% | 6,659,104 | 2 | OHIO | 108,280 | 8.01% | 1,239,811 |
| 3 | ALABAMA | 391,953 | 11.13% | 5,867,541 | 3 | INDIANA | 96,954 | 7.18% | 1,110,124 |
| 4 | MISSISSIPPI | 330,982 | 9.40% | 4,954,799 | 4 | PENNSYLVANIA | 87,930 | 6.51% | 1,006,794 |
| 5 | NORTH CAROLINA | 329,498 | 9.36% | 4,932,592 | 5 | CALIFORNIA | 84,367 | 6.24% | 965,997 |
| 6 | TEXAS | 260,686 | 7.40% | 3,902,473 | 6 | GEORGIA | 77,093 | 5.71% | 882,712 |
| 7 | MARYLAND | 143,964 | 4.09% | 2,155,138 | 7 | TEXAS | 76,467 | 5.66% | 875,545 |
| 8 | DELAWARE | 112,291 | 3.19% | 1,680,999 | 8 | ARKANSAS | 55,911 | 4.14% | 640,183 |
| 9 | KENTUCKY | 109,399 | 3.11% | 1,637,707 | 9 | NORTH CAROLINA | 50,993 | 3.77% | 583,871 |
| 10 | MISSOURI | 102,537 | 2.91% | 1,534,984 | 10 | FLORIDA | 47,151 | 3.49% | 539,879 |
| 11 | SOUTH CAROLINA | 100,642 | 2.86% | 1,506,618 | 11 | MINNESOTA | 42,386 | 3.14% | 485,323 |
| 12 | CALIFORNIA | 97,548 | 2.77% | 1,460,290 | 12 | NEBRASKA | 41,950 | 3.10% | 480,326 |
| 13 | OKLAHOMA | 97,395 | 2.77% | 1,458,000 | 13 | ALABAMA | 38,497 | 2.85% | 440,791 |
| 14 | VIRGINIA | 96,142 | 2.73% | 1,439,247 | 14 | MICHIGAN | 36,137 | 2.67% | 413,773 |
| 15 | TENNESSEE | 90,198 | 2.56% | 1,350,271 | 15 | MISSOURI | 28,998 | 2.15% | 332,023 |
| 16 | LOUISIANA | 79,750 | 2.26% | 1,193,850 | 16 | MISSISSIPPI | 24,948 | 1.85% | 285,652 |
| 17 | PENNSYLVANIA | 60,459 | 1.72% | 905,067 | 17 | WASHINGTON | 23,143 | 1.71% | 264,983 |
| 18 | FLORIDA | 31,041 | 0.88% | 464,685 | 18 | ILLINOIS | 21,142 | 1.56% | 242,080 |
| 19 | WEST VIRGINIA | 28,162 | 0.80% | 421,581 | 19 | WISCONSIN | 19,495 | 1.44% | 223,214 |
| 20 | OHIO | 22,026 | 0.63% | 329,733 | 20 | SOUTH CAROLINA | 18,857 | 1.40% | 215,917 |
| 21 | MINNESOTA | 19,010 | 0.54% | 284,580 | 21 | KENTUCKY | 18,338 | 1.36% | 209,972 |
| 22 | WISCONSIN | 15,517 | 0.44% | 232,292 | 22 | NEW YORK | 15,812 | 1.17% | 181,046 |
| 23 | INDIANA | 12,169 | 0.35% | 182,171 | 23 | COLORADO | 15,612 | 1.16% | 178,755 |
| 24 | WASHINGTON | 10,214 | 0.29% | 152,899 | 24 | UTAH | 14,339 | 1.06% | 164,183 |
| 25 | OREGON | 8,804 | 0.25% | 131,799 | 25 | OKLAHOMA | 13,295 | 0.98% | 152,230 |
| 26 | IOWA | 3,964 | 0.11% | 59,335 | 26 | VIRGINIA | 12,836 | 0.95% | 146,968 |
| 27 | NEBRASKA | 1,699 | 0.05% | 25,431 | 27 | SOUTH DAKOTA | 11,683 | 0.86% | 133,773 |
| 28 | MICHIGAN | 1,500 | 0.04% | 22,448 | 28 | OREGON | 10,946 | 0.81% | 125,330 |
| 29 | NEW YORK | 1,031 | 0.03% | 15,429 | 29 | MARYLAND | 10,651 | 0.79% | 121,953 |
| 30 | ILLINOIS | 239 | 0.01% | 3,584 | 30 | LOUISIANA | 7,968 | 0.59% | 91,231 |
| 31 | MONTANA | 237 | 0.01% | 3,552 | 31 | TENNESSEE | 6,854 | 0.51% | 78,473 |
| 32 | SOUTH DAKOTA | 225 | 0.01% | 3,363 | 32 | NEW JERSEY | 6,241 | 0.46% | 71,456 |
| 33 | CONNECTICUT | 221 | 0.01% | 3,308 | 33 | WEST VIRGINIA | 4,881 | 0.36% | 55,889 |
| 34 | VERMONT | 93 | 0% | 1,398 | 34 | HAWAII | 1,473 | 0.11% | 16,865 |
| 35 | NEW HAMPSHIRE | 53 | 0% | 796 | 35 | MONTANA | 1,421 | 0.11% | 16,269 |
| 36 | KANSAS | 43 | 0% | 643 | 36 | VERMONT | 894 | 0.07% | 10,241 |
| 37 | NEW JERSEY | 39 | 0% | 589 | 37 | NEW HAMPSHIRE | 842 | 0.06% | 9,635 |
| 38 | NORTH DAKOTA | 35 | 0% | 520 | 38 | MASSACHUSETTS | 559 | 0.04% | 6,401 |
| 39 | MAINE | 33 | 0% | 489 | 39 | NORTH DAKOTA | 437 | 0.03% | 5,008 |
| 40 | NEW MEXICO | 25 | 0% | 369 | 40 | RHODE ISLAND | 183 | 0.01% | 2,099 |
| 41 | COLORADO | 24 | 0% | 364 | 41 | WYOMING | 65 | 0% | 744 |
| 42 | IDAHO | 17 | 0% | 261 | 42 | NEVADA | 23 | 0% | 268 |
| 43 | UTAH | 6 | 0% | 84 | 43 | ALASKA | 14 | 0% | 166 |
| 44 | HAWAII | 5 | 0% | 70 | 44 | ARIZONA | D | | |
| 45 | ALASKA | 5 | 0% | 69 | 45 | CONNECTICUT | D | | |
| 46 | ARIZONA | 4 | 0% | 66 | 46 | DELAWARE | D | | |
| 47 | WYOMING | 3 | 0% | 50 | 47 | IDAHO | D | | |
| 48 | NEVADA | 1 | 0% | 10 | 48 | KANSAS | D | | |
| 49 | MASSACHUSETTS | D | | | 49 | MAINE | D | | |
| 50 | RHODE ISLAND | D | | | 50 | NEW MEXICO | D | | |
| | U.S. TOTAL | 3,522,083 | | 52,725,576 | | U.S. TOTAL | 1,351,241 | | 15,471,706 |

*"D" signifies that the data were not disclosed (because there were too few producers in the category to protect confidentiality).*

*USDA 2009a.*

**Table A-7. Total animal units and estimated tons of manure produced for turkeys, as well as all poultry (broilers, layers, and turkeys combined) in 2007.**

| National Rank | State | Total Turkey AUs | Percent of Total Turkey AUs | Total Tons Manure | National Rank | State | Total Poultry AUs | Percent of Total Poultry AUs | Total Tons Manure |
|---|---|---|---|---|---|---|---|---|---|
| 1 | MINNESOTA | 273,109 | 17.41% | 2,234,033 | 1 | NORTH CAROLINA | 647,147 | 10.05% | 7,697,703 |
| 2 | NORTH CAROLINA | 266,655 | 17.00% | 2,181,239 | 2 | ARKANSAS | 641,595 | 9.96% | 8,451,469 |
| 3 | ARKANSAS | 140,853 | 8.98% | 1,152,181 | 3 | GEORGIA | 594,486 | 9.23% | 8,627,880 |
| 4 | MISSOURI | 128,421 | 8.19% | 1,050,486 | 4 | ALABAMA | 430,581 | 6.68% | 6,309,404 |
| 5 | CALIFORNIA | 100,048 | 6.38% | 818,394 | 5 | TEXAS | 366,807 | 5.69% | 5,020,588 |
| 6 | VIRGINIA | 94,492 | 6.02% | 772,944 | 6 | MISSISSIPPI | 355,951 | 5.53% | 5,240,622 |
| 7 | INDIANA | 89,128 | 5.68% | 729,064 | 7 | MINNESOTA | 334,506 | 5.19% | 3,003,937 |
| 8 | SOUTH CAROLINA | 81,854 | 5.22% | 669,564 | 8 | CALIFORNIA | 281,962 | 4.38% | 3,244,681 |
| 9 | IOWA | 59,733 | 3.81% | 488,616 | 9 | IOWA | 278,871 | 4.33% | 3,011,703 |
| 10 | WISCONSIN | 55,010 | 3.51% | 449,979 | 10 | MISSOURI | 259,956 | 4.04% | 2,917,493 |
| 11 | PENNSYLVANIA | 52,799 | 3.37% | 431,894 | 11 | VIRGINIA | 203,470 | 3.16% | 2,359,159 |
| 12 | SOUTH DAKOTA | 33,322 | 2.12% | 272,574 | 12 | SOUTH CAROLINA | 201,354 | 3.13% | 2,392,098 |
| 13 | UTAH | 32,676 | 2.08% | 267,293 | 13 | PENNSYLVANIA | 201,187 | 3.12% | 2,343,756 |
| 14 | OHIO | 30,966 | 1.97% | 253,305 | 14 | INDIANA | 198,251 | 3.08% | 2,021,359 |
| 15 | TEXAS | 29,654 | 1.89% | 242,569 | 15 | OHIO | 161,273 | 2.50% | 1,822,849 |
| 16 | MICHIGAN | 29,535 | 1.88% | 241,599 | 16 | MARYLAND | 157,947 | 2.45% | 2,304,346 |
| 17 | WEST VIRGINIA | 24,494 | 1.56% | 200,364 | 17 | KENTUCKY | 128,197 | 1.99% | 1,851,437 |
| 18 | ILLINOIS | 12,626 | 0.80% | 103,284 | 18 | DELAWARE | 112,302 | 1.74% | 1,681,085 |
| 19 | NEBRASKA | 11,362 | 0.72% | 92,938 | 19 | OKLAHOMA | 110,690 | 1.72% | 1,610,230 |
| 20 | KANSAS | 8,380 | 0.53% | 68,551 | 20 | TENNESSEE | 97,104 | 1.51% | 1,429,169 |
| 21 | NORTH DAKOTA | 6,631 | 0.42% | 54,241 | 21 | WISCONSIN | 90,022 | 1.40% | 905,486 |
| 22 | MARYLAND | 3,332 | 0.21% | 27,254 | 22 | LOUISIANA | 87,729 | 1.36% | 1,285,179 |
| 23 | NEW YORK | 1,483 | 0.09% | 12,128 | 23 | FLORIDA | 78,398 | 1.22% | 1,006,247 |
| 24 | KENTUCKY | 459 | 0.03% | 3,759 | 24 | MICHIGAN | 67,172 | 1.04% | 677,820 |
| 25 | NEW JERSEY | 275 | 0.02% | 2,247 | 25 | WEST VIRGINIA | 57,537 | 0.89% | 677,834 |
| 26 | MASSACHUSETTS | 261 | 0.02% | 2,137 | 26 | NEBRASKA | 55,010 | 0.85% | 598,696 |
| 27 | MONTANA | 243 | 0.02% | 1,990 | 27 | UTAH | 47,021 | 0.73% | 431,561 |
| 28 | FLORIDA | 206 | 0.01% | 1,682 | 28 | SOUTH DAKOTA | 45,230 | 0.70% | 409,710 |
| 29 | ALABAMA | 131 | 0.01% | 1,073 | 29 | ILLINOIS | 34,008 | 0.53% | 348,948 |
| 30 | NEW MEXICO | 92 | 0.01% | 752 | 30 | WASHINGTON | 33,413 | 0.52% | 418,344 |
| 31 | VERMONT | 86 | 0.01% | 702 | 31 | OREGON | 19,795 | 0.31% | 257,497 |
| 32 | WASHINGTON | 57 | 0.00% | 463 | 32 | NEW YORK | 18,325 | 0.28% | 208,603 |
| 33 | CONNECTICUT | 53 | 0.00% | 435 | 33 | COLORADO | 15,636 | 0.24% | 179,119 |
| 34 | TENNESSEE | 52 | 0.00% | 425 | 34 | KANSAS | 8,423 | 0.13% | 69,194 |
| 35 | MAINE | 46 | 0.00% | 378 | 35 | NORTH DAKOTA | 7,103 | 0.11% | 59,769 |
| 36 | OREGON | 45 | 0.00% | 369 | 36 | NEW JERSEY | 6,555 | 0.10% | 74,293 |
| 37 | NEW HAMPSHIRE | 38 | 0.00% | 309 | 37 | MONTANA | 1,901 | 0.03% | 21,811 |
| 38 | GEORGIA | 30 | 0.00% | 242 | 38 | HAWAII | 1,479 | 0.02% | 16,947 |
| 39 | RHODE ISLAND | 29 | 0.00% | 233 | 39 | VERMONT | 1,074 | 0.02% | 12,341 |
| 40 | MISSISSIPPI | 21 | 0.00% | 170 | 40 | NEW HAMPSHIRE | 933 | 0.01% | 10,741 |
| 41 | IDAHO | 19 | 0.00% | 152 | 41 | MASSACHUSETTS | 820 | 0.01% | 8,538 |
| 42 | ARIZONA | 13 | 0.00% | 105 | 42 | CONNECTICUT | 274 | 0.00% | 3,743 |
| 43 | LOUISIANA | 12 | 0.00% | 98 | 43 | RHODE ISLAND | 212 | 0.00% | 2,332 |
| 44 | ALASKA | 11 | 0.00% | 88 | 44 | NEW MEXICO | 117 | 0.00% | 1,121 |
| 45 | DELAWARE | 10 | 0.00% | 86 | 45 | MAINE | 79 | 0.00% | 867 |
| 46 | WYOMING | 7 | 0.00% | 54 | 46 | WYOMING | 75 | 0.00% | 848 |
| 47 | NEVADA | 2 | 0.00% | 18 | 47 | IDAHO | 36 | 0.00% | 412 |
| 48 | HAWAII | 1 | 0.00% | 12 | 48 | ALASKA | 30 | 0.00% | 323 |
| 49 | COLORADO | D | | | 49 | NEVADA | 26 | 0.00% | 296 |
| 50 | OKLAHOMA | D | | | 50 | ARIZONA | 17 | 0.00% | 171 |
| | U.S. TOTAL | 1,568,762 | | 12,832,472 | | U.S. TOTAL | 6,442,085 | | 81,029,754 |

*"D" signifies that the data were not disclosed (because there were too few producers in the category to protect confidentiality).*
*USDA 2009a.*

## Table A-8. Livestock (cattle, swine, and poultry) animal units as a total and per acre of farmland in 2007.

| Rank Total AUs | State | AUs | Rank AUs/Acre Farmland | State | AUs/Acre Farmland |
|---|---|---|---|---|---|
| 1 | TEXAS | 11,109,770 | 1 | NORTH CAROLINA | 0.32 |
| 2 | IOWA | 5,586,515 | 2 | DELAWARE | 0.26 |
| 3 | NEBRASKA | 5,235,899 | 3 | PENNSYLVANIA | 0.23 |
| 4 | CALIFORNIA | 5,235,439 | 4 | VERMONT | 0.22 |
| 5 | KANSAS | 4,932,902 | 5 | WISCONSIN | 0.21 |
| 6 | OKLAHOMA | 4,571,012 | 6 | CALIFORNIA | 0.21 |
| 7 | MISSOURI | 4,178,962 | 7 | NEW YORK | 0.20 |
| 8 | MINNESOTA | 3,268,570 | 8 | VIRGINIA | 0.19 |
| 9 | WISCONSIN | 3,213,092 | 9 | IOWA | 0.18 |
| 10 | SOUTH DAKOTA | 3,179,772 | 10 | TENNESSEE | 0.17 |
| 11 | NORTH CAROLINA | 2,704,347 | 11 | FLORIDA | 0.17 |
| 12 | COLORADO | 2,183,438 | 12 | IDAHO | 0.17 |
| 13 | MONTANA | 2,172,114 | 13 | MARYLAND | 0.16 |
| 14 | ARKANSAS | 2,164,456 | 14 | ALABAMA | 0.16 |
| 15 | KENTUCKY | 2,143,425 | 15 | ARKANSAS | 0.16 |
| 16 | IDAHO | 1,934,024 | 16 | GEORGIA | 0.16 |
| 17 | TENNESSEE | 1,901,829 | 17 | KENTUCKY | 0.15 |
| 18 | PENNSYLVANIA | 1,801,172 | 18 | MISSOURI | 0.14 |
| 19 | ILLINOIS | 1,623,316 | 19 | OKLAHOMA | 0.13 |
| 20 | GEORGIA | 1,582,445 | 20 | MINNESOTA | 0.12 |
| 21 | FLORIDA | 1,572,198 | 21 | CONNECTICUT | 0.12 |
| 22 | VIRGINIA | 1,540,785 | 22 | SOUTH CAROLINA | 0.12 |
| 23 | NORTH DAKOTA | 1,511,460 | 23 | HAWAII | 0.12 |
| 24 | OHIO | 1,486,479 | 24 | NEBRASKA | 0.12 |
| 25 | ALABAMA | 1,446,018 | 25 | MICHIGAN | 0.11 |
| 26 | INDIANA | 1,426,494 | 26 | WEST VIRGINIA | 0.11 |
| 27 | NEW YORK | 1,405,612 | 27 | OHIO | 0.11 |
| 28 | NEW MEXICO | 1,369,823 | 28 | KANSAS | 0.11 |
| 29 | OREGON | 1,182,494 | 29 | LOUISIANA | 0.10 |
| 30 | MISSISSIPPI | 1,171,555 | 30 | MISSISSIPPI | 0.10 |
| 31 | MICHIGAN | 1,136,740 | 31 | INDIANA | 0.10 |
| 32 | WYOMING | 1,101,102 | 32 | MASSACHUSETTS | 0.09 |
| 33 | WASHINGTON | 980,293 | 33 | TEXAS | 0.09 |
| 34 | LOUISIANA | 840,534 | 34 | RHODE ISLAND | 0.08 |
| 35 | ARIZONA | 813,626 | 35 | NEW HAMPSHIRE | 0.08 |
| 36 | UTAH | 760,972 | 36 | SOUTH DAKOTA | 0.07 |
| 37 | SOUTH CAROLINA | 586,368 | 37 | OREGON | 0.07 |
| 38 | WEST VIRGINIA | 394,816 | 38 | COLORADO | 0.07 |
| 39 | NEVADA | 380,843 | 39 | UTAH | 0.07 |
| 40 | MARYLAND | 333,955 | 40 | WASHINGTON | 0.07 |
| 41 | VERMONT | 268,767 | 41 | NEVADA | 0.06 |
| 42 | DELAWARE | 132,875 | 42 | MAINE | 0.06 |
| 43 | HAWAII | 130,823 | 43 | ILLINOIS | 0.06 |
| 44 | MAINE | 82,780 | 44 | NEW JERSEY | 0.06 |
| 45 | CONNECTICUT | 48,862 | 45 | NORTH DAKOTA | 0.04 |
| 46 | MASSACHUSETTS | 45,418 | 46 | WYOMING | 0.04 |
| 47 | NEW JERSEY | 41,652 | 47 | MONTANA | 0.04 |
| 48 | NEW HAMPSHIRE | 36,402 | 48 | NEW MEXICO | 0.03 |
| 49 | ALASKA | 11,903 | 49 | ARIZONA | 0.03 |
| 50 | RHODE ISLAND | 5,364 | 50 | ALASKA | 0.01 |
|  | U.S. TOTAL | 92,969,509 |  | U.S. AVERAGE | 0.12 |

*USDA 2009a.*

**Table A-9. Total estimated livestock and poultry (cattle, swine, and poultry) manure and estimated tons of manure per acre of farmland in 2007.**

| Rank Total Manure | State | Tons Manure | Rank Tons Manure/Acre Farmland | State | Tons Manure/Acre Farmland |
|---|---|---|---|---|---|
| 1 | TEXAS | 128,048,896 | 1 | NORTH CAROLINA | 3.85 |
| 2 | CALIFORNIA | 68,496,143 | 2 | DELAWARE | 3.81 |
| 3 | IOWA | 68,360,493 | 3 | VERMONT | 3.05 |
| 4 | NEBRASKA | 59,100,556 | 4 | PENNSYLVANIA | 2.99 |
| 5 | KANSAS | 55,792,510 | 5 | WISCONSIN | 2.80 |
| 6 | OKLAHOMA | 52,036,892 | 6 | CALIFORNIA | 2.70 |
| 7 | MISSOURI | 48,070,611 | 7 | NEW YORK | 2.66 |
| 8 | WISCONSIN | 42,531,594 | 8 | MARYLAND | 2.23 |
| 9 | MINNESOTA | 39,816,914 | 9 | VIRGINIA | 2.22 |
| 10 | SOUTH DAKOTA | 36,358,712 | 10 | IOWA | 2.22 |
| 11 | NORTH CAROLINA | 32,620,857 | 11 | IDAHO | 2.13 |
| 12 | ARKANSAS | 25,665,769 | 12 | TENNESSEE | 2.02 |
| 13 | KENTUCKY | 25,117,706 | 13 | FLORIDA | 2.01 |
| 14 | COLORADO | 25,022,958 | 14 | GEORGIA | 1.99 |
| 15 | MONTANA | 24,709,841 | 15 | ALABAMA | 1.98 |
| 16 | IDAHO | 24,532,956 | 16 | ARKANSAS | 1.85 |
| 17 | PENNSYLVANIA | 23,359,900 | 17 | KENTUCKY | 1.80 |
| 18 | TENNESSEE | 22,209,901 | 18 | MISSOURI | 1.66 |
| 19 | GEORGIA | 20,202,017 | 19 | CONNECTICUT | 1.62 |
| 20 | ILLINOIS | 19,190,293 | 20 | OKLAHOMA | 1.48 |
| 21 | NEW YORK | 19,084,031 | 21 | MINNESOTA | 1.48 |
| 22 | FLORIDA | 18,563,926 | 22 | MICHIGAN | 1.46 |
| 23 | OHIO | 18,460,395 | 23 | SOUTH CAROLINA | 1.41 |
| 24 | VIRGINIA | 18,029,169 | 24 | HAWAII | 1.33 |
| 25 | ALABAMA | 17,902,968 | 25 | OHIO | 1.32 |
| 26 | INDIANA | 17,368,868 | 26 | NEBRASKA | 1.30 |
| 27 | NORTH DAKOTA | 17,175,740 | 27 | MISSISSIPPI | 1.27 |
| 28 | NEW MEXICO | 17,173,686 | 28 | WEST VIRGINIA | 1.23 |
| 29 | MICHIGAN | 14,628,016 | 29 | LOUISIANA | 1.23 |
| 30 | MISSISSIPPI | 14,583,109 | 30 | KANSAS | 1.20 |
| 31 | OREGON | 13,989,238 | 31 | INDIANA | 1.18 |
| 32 | WYOMING | 12,472,771 | 32 | MASSACHUSETTS | 1.14 |
| 33 | WASHINGTON | 12,340,841 | 33 | NEW HAMPSHIRE | 1.03 |
| 34 | ARIZONA | 10,074,405 | 34 | RHODE ISLAND | 0.99 |
| 35 | LOUISIANA | 9,968,209 | 35 | TEXAS | 0.98 |
| 36 | UTAH | 8,938,890 | 36 | OREGON | 0.85 |
| 37 | SOUTH CAROLINA | 6,895,155 | 37 | SOUTH DAKOTA | 0.83 |
| 38 | MARYLAND | 4,578,248 | 38 | WASHINGTON | 0.82 |
| 39 | WEST VIRGINIA | 4,548,748 | 39 | MAINE | 0.82 |
| 40 | NEVADA | 4,459,013 | 40 | UTAH | 0.81 |
| 41 | VERMONT | 3,757,488 | 41 | COLORADO | 0.79 |
| 42 | DELAWARE | 1,944,044 | 42 | NEVADA | 0.76 |
| 43 | HAWAII | 1,490,546 | 43 | ILLINOIS | 0.72 |
| 44 | MAINE | 1,100,800 | 44 | NEW JERSEY | 0.71 |
| 45 | CONNECTICUT | 658,068 | 45 | NORTH DAKOTA | 0.43 |
| 46 | MASSACHUSETTS | 588,513 | 46 | WYOMING | 0.41 |
| 47 | NEW JERSEY | 521,513 | 47 | MONTANA | 0.40 |
| 48 | NEW HAMPSHIRE | 486,181 | 48 | NEW MEXICO | 0.40 |
| 49 | ALASKA | 137,131 | 49 | ARIZONA | 0.39 |
| 50 | RHODE ISLAND | 67,158 | 50 | ALASKA | 0.16 |
|  | U.S. TOTAL | 1,113,232,385 |  | U.S. AVERAGE | 1.50 |

*USDA 2009a.*

## Table A-10. Freshwater and saltwater aquaculture farms in the U.S. during 2005.

| Geographic Area | Rank Farms | Freshwater Farms | Saltwater Farms | Total Farms |
|---|---|---|---|---|
| Louisiana | 1 | 738 | 135 | 873 |
| Mississippi | 2 | 403 | 1 | 403 |
| Florida | 3 | 196 | 163 | 359 |
| Alabama | 4 | 213 | 2 | 215 |
| Arkansas | 5 | 211 | - | 211 |
| Washington | 6 | 21 | 175 | 194 |
| North Carolina | 7 | 129 | 57 | 186 |
| Massachusetts | 8 | 18 | 140 | 157 |
| Virginia | 9 | 28 | 122 | 147 |
| California | 10 | 96 | 22 | 118 |
| Texas | 11 | 79 | 19 | 95 |
| New Jersey | 12 | 17 | 70 | 87 |
| Maryland | 13 | 11 | 75 | 86 |
| South Carolina | 14 | 43 | 45 | 85 |
| Wisconsin | 15 | 84 | - | 84 |
| Georgia | 16 | 78 | 1 | 79 |
| Minnesota | 17 | 77 | - | 77 |
| Kentucky | 18 | 65 | - | 65 |
| Hawaii | 19 | 33 | 30 | 59 |
| Pennsylvania | 20 | 56 | - | 56 |
| Ohio | 21 | 55 | - | 55 |
| New York | 22 | 41 | 13 | 54 |
| Maine | 23 | 10 | 40 | 50 |
| Oregon | 24 | 26 | 21 | 47 |
| Illinois | 24 | 47 | 1 | 47 |
| Tennessee | 26 | 45 | - | 45 |
| Missouri | 27 | 35 | - | 35 |
| Idaho | 27 | 35 | - | 35 |
| Michigan | 29 | 34 | 1 | 34 |
| Connecticut | 30 | 3 | 27 | 30 |
| Nebraska | 31 | 26 | - | 26 |
| Alaska | 31 | 1 | 25 | 26 |
| Iowa | 33 | 21 | - | 21 |
| West Virginia | 33 | 21 | - | 21 |
| Oklahoma | 35 | 20 | - | 20 |
| Indiana | 36 | 17 | 1 | 18 |
| Colorado | 37 | 15 | - | 15 |
| Kansas | 38 | 12 | - | 12 |
| Rhode Island | 38 | 2 | 11 | 12 |
| Utah | 40 | 11 | - | 11 |
| Arizona | 40 | 11 | - | 11 |
| New Hampshire | 42 | 5 | 6 | 10 |
| Vermont | 43 | 9 | - | 9 |
| Montana | 44 | 8 | - | 8 |
| South Dakota | 45 | 7 | - | 7 |
| Wyoming | 45 | 7 | - | 7 |
| New Mexico | 47 | 3 | - | 3 |
| Delaware | 47 | 3 | - | 3 |
| North Dakota | 49 | 1 | - | 1 |
| Nevada | 50 | - | - | - |
| **United States** | | **3,127** | **1,203** | **4,309** |

*USDA 2006.*

**Table A-11. Aquaculture in the U.S. presented as total acres and sales.**

| Geographic Area | Rank Acres | Freshwater Acres | Saltwater Acres | Total Acres | Total Sales (1,000s of $) | Rank $ |
|---|---|---|---|---|---|---|
| Louisiana | 1 | 104,645 | 215,770 | 320,415 | $101,314 | 4 |
| Mississippi | 2 | 102,898 | D | 102,898 | $249,704 | 1 |
| Connecticut | 3 | D | 62,959 | 62,959 | $12,902 | 14 |
| Arkansas | 4 | 61,135 | - | 61,135 | $110,542 | 2 |
| Minnesota | 5 | 41,023 | - | 41,023 | $8,412 | 19 |
| Alabama | 6 | 25,351 | D | 25,351 | $102,796 | 3 |
| Washington | 7 | 209 | 13,269 | 13,478 | $93,203 | 5 |
| Virginia | 8 | 143 | 12,412 | 12,555 | $40,939 | 8 |
| California | 9 | 3,338 | 6,002 | 9,340 | $69,607 | 6 |
| Texas | 10 | 4,651 | 2,432 | 7,083 | $35,359 | 10 |
| New Jersey | 11 | 51 | 4,466 | 4,517 | $3,714 | 25 |
| North Carolina | 12 | 3,463 | 707 | 4,170 | $24,725 | 12 |
| Florida | 13 | 2,292 | 718 | 3,010 | $57,406 | 7 |
| Missouri | 14 | 2,689 | - | 2,689 | $7,144 | 22 |
| Oregon | 15 | 101 | 2,425 | 2,526 | $12,478 | 15 |
| South Carolina | 16 | 683 | 1,531 | 2,214 | $4,773 | 24 |
| Wisconsin | 17 | 1,977 | - | 1,977 | $7,025 | 23 |
| Georgia | 18 | 1,914 | D | 1,914 | $7,502 | 20 |
| Massachusetts | 19 | 60 | 1,108 | 1,168 | $9,342 | 16 |
| South Dakota | 20 | 1,066 | - | 1,066 | $484 | 42 |
| Illinois | 21 | 805 | D | 805 | $3,176 | 28 |
| Ohio | 22 | 759 | - | 759 | $3,185 | 27 |
| Tennessee | 23 | 707 | - | 707 | $1,286 | 35 |
| Pennsylvania | 24 | 626 | - | 626 | $8,951 | 17 |
| Kentucky | 25 | 624 | - | 624 | $2,341 | 30 |
| Maine | 26 | 32 | 585 | 617 | $25,580 | 11 |
| Iowa | 27 | 594 | - | 594 | $1,469 | 34 |
| Kansas | 28 | 590 | - | 590 | $342 | 43 |
| Oklahoma | 29 | 557 | - | 557 | $1,958 | 31 |
| Nebraska | 30 | 503 | - | 503 | $1,750 | 33 |
| Indiana | 31 | 443 | D | 443 | D | - |
| Michigan | 32 | 429 | D | 429 | $2,398 | 29 |
| Maryland | 33 | 155 | 238 | 393 | $7,292 | 21 |
| New York | 34 | 385 | D | 385 | $8,913 | 18 |
| Hawaii | 35 | 75 | 254 | 329 | $13,761 | 13 |
| Idaho | 36 | 151 | - | 151 | $37,685 | 9 |
| Alaska | 37 | D | 148 | 148 | $826 | 39 |
| Colorado | 38 | 85 | - | 85 | $3,349 | 26 |
| New Hampshire | 39 | 10 | 70 | 80 | $1,054 | 37 |
| Rhode Island | 40 | D | 51 | 51 | $840 | 38 |
| West Virginia | 41 | 48 | - | 48 | $1,145 | 36 |
| Utah | 42 | 38 | - | 38 | $559 | 41 |
| Wyoming | 43 | 37 | - | 37 | $209 | 45 |
| Arizona | 44 | 31 | - | 31 | $562 | 40 |
| Montana | 45 | 13 | - | 13 | $302 | 44 |
| Vermont | 46 | 11 | - | 11 | $80 | 46 |
| New Mexico | 47 | 1 | - | 1 | D | - |
| Delaware | - | D | - | - | $1,870 | 32 |
| North Dakota | - | D | - | - | D | - |
| Nevada | - | - | - | - | - | - |
| **United States** | | **365,566** | **327,487** | **693,053** | **$1,092,386** | |

*"D" signifies that the data were not disclosed (because there were too few producers in the category to protect confidentiality).*
*USDA 2006.*

# Appendix 2. Animal Life Stages

**Table A-12. Livestock animal type and life stages definitions.**

| Animal Type | Term | Definition |
|---|---|---|
| Cattle (Beef, Dairy) | Bovine | General term for cattle |
| | Dairy Cow | A female cow that produces milk for human consumption, or raises replacement heifers |
| | Heifer | A female cow that has not yet had her first calf. Typically less than three years of age |
| | Beef Cattle | Cattle raised for meat production |
| | Steer | A castrated bovine male |
| | Calf | A male or female bovine under one year of age |
| | Preweaned Calf | Calves that are nursing from their mother or a dam (i.e., a female parent in pedigree) |
| | Weaned Heifer | Heifers that are no longer nursing |
| | Replacement Heifer | Cows raised to replace those currently in the herd |
| | Lactating | A cow that is producing milk |
| | Non-Lactating | A cow that is dry (i.e., not secreting milk). Cows are typically provided a dry period between lactations to allow the cow's udders an opportunity to regenerate secretory tissue |
| | Cow-Calf Operation | A facility that maintains breeding bovine and produces weaned calves |
| | Growing | A cattle grown to market weight |
| | Feedlot | Beef cattle in confined, outdoor pens and fed a high-energy ration of grains and other concentrates |
| Swine | Hog | General term for growing swine |
| | Sow | A female after she has borne a litter |
| | Farrow | The life stage between birth and weaning |
| | Preweaned | Pigs that are still nursing and have not yet been removed from the sow |
| | Nursery (Weaned) | Pigs that are no longer nursing and have been removed from the sow |
| | Breeder | Swine that produce offspring |
| | Grower/Feeder/Finisher/Market | Swine that are fed until they reach market weight and are ready for slaughter |
| Poultry (Broilers, Layers, and Turkeys) | Broiler | A chicken utilized for meat production |
| | Layer | A chicken utilized for egg production |
| | Pullet | A laying hen prior to laying its first egg |
| | Grower/Finisher | Birds grown to market weight and sent to slaughter |
| | Breeder | A bird that produces offspring |

*MacDonald and McBride 2009 and USEPA 2009c.*

This page intentionally left blank.

# Appendix 3. Additional Technical Resources for Manure Management

This appendix includes a sampling of on-line resources that are available to help planners and producers with manure management. It is intended to illustrate the breadth of information available and identify the agencies and organizations that are working actively to provide information to planners and producers.

## World Health Organization

- Animal Waste Water Quality and Human Health- This website provides links to resources published by WHO on water sanitation and health including a book which contains relevant information in connection with pathogens on the scope of domestic animal manure discharge into the environment, the fate and transport of the discharged manure and the pathogens they may contain, human exposure to the manure, potential health effects associated with those exposures and interventions that can limit human exposures to livestock manure. It also addresses the monitoring, detection and effectiveness of the best management practices related to these issues.
  http://www.who.int/water_sanitation_health/publications/2012/animal_waste/en/

## U.S. Department of Agriculture

- USDAs NRCS Technical Standard 590 Nutrient Management- This website contains the USDAs 590 Nutrient Management Conservation Practice Standard. The standard provides guidance on managing nutrient applications to meet crop needs and minimize nonpoint source pollution. http://www.nrcs.usda.gov/Internet/FSE_DOCUMENTS/stelprdb1046433.pdf

- USDAs Manure and Nutrient Management Resources- This site contains many helpful resources including a guide that provides a complete review of key management practices and methods to minimize waste pollution, software that may be used by large livestock and poultry facility operators and owners to estimate manure generation and production of process water, and training courses on water quality, waste management, nutrient and pest management, conservation practices, and planning and designing animal waste containment.
  http://go.usa.gov/Mo

- USDAs ERS Manure Management Website- This website is an important resource for publications and economic research related to animal and manure production. http://www.ers.usda.gov/browse/view.aspx?subtitle=Practices+Management+Manure+Management

- USDAs Agricultural Research Service (ARS) Website- The ARS has ongoing efforts designed to enhance current practices and develop new methods for efficiently and effectively managing manure. http://www.ars.usda.gov/research/programs/programs.htm?NP_CODE=21

## U.S. Environmental Protection Agency

- The USEPAs Agricultural Center Website- This is the USEPAs primary website for agricultural planning, management and needs.
  http://www.epa.gov/agriculture

- Animal Feeding Operations (AFO) Virtual Information Center- The Animal Feeding Operations (AFO) Virtual Information Center is a tool to facilitate quick access to livestock and poultry agricultural

information in the U.S. This site is a single point of reference to obtain links to state regulations, web sites, permits and policies, nutrient management information, livestock and trade associations, federal web sites, best management practices and controls, cooperative extension and land grant universities, research funding, and information on environmental issues.
http://cfpub.epa.gov/npdes/afo/virtualcenter.cfm

- National Management Measures to Control Nonpoint Source Pollution from Agriculture- This guidance document contains economically achievable best management practices designed to reduce agricultural pollution to surface and ground water. http://water.epa.gov/polwaste/nps/agriculture/agmm_index.cfm

- The USEPA's Nonpoint Source Pollution Publications & Resources Website- This website provides links and references to nonpoint source materials for both professionals and the public. http://water.epa.gov/polwaste/nps/pubs.cfm

- The USEPA's Source Water Protection Program - This website provides information and resources about protecting surface water and ground water drinking water sources. http://water.epa.gov/infrastructure/drinkingwater/sourcewater/protection/index.cfm

- Healthy Watersheds Initiative- This website provides information on the concept and benefits of protecting healthy unimpaired waters from degradation and also provides information on conservation approaches and tools.
http://water.epa.gov/polwaste/nps/watershed/index.cfm

- EPA's Nutrient Indicators Dataset- The Dataset consists of a set of nine indicators and associated state-level data to serve as a regional compendium of information pertaining to documented or potential nitrogen and phosphorus pollution, impacts of this pollution, and states' efforts to minimize loadings and adopt nutrient criteria.
www.epa.gov/nutrientpollution/Dataset

## Additional State and University Technical Resources

The list of useful resources available from state and university resources is too lengthy to include in this report. In addition to the exceptional resources listed below, please contact applicable university extension services and state agencies responsible for natural resources management and environmental protection for more information.

- eXtension- This website provides objective and research-based information and learning opportunities that help people improve their lives. eXtension is an educational partnership of 74 universities in the U.S.
http://www.extension.org/animal_manure_management

- Cornell University's Cornell Dairy Environmental Systems Program - This program provides information to dairy farmers to help manage their businesses in a way that protects the environment. This program also focuses on renewable energy (dairy manure-based anaerobic digestion).
http://www.manuremanagement.cornell.edu/

- Ohio State University's Ohio Composting and Manure Management Website- This program researches, develops and communicates sustainable strategies for management of animal manure and nutrients. Resources provided include workshops and literature on topics such as composting, application of liquid manure, and ammonia emissions and nitrogen conservation.
http://www.oardc.ohio-state.edu/ocamm/t01page/iewome.htm

- The University of Illinois Manure Central Website- This website directs the reader to a variety of resources for topics such as composting, manure management plans, a manure exchange program, and manure management forms all farms.
http://web.extension.illinois.edu/manurecentral

- Wisconsin Manure Management Advisory System - This website provides tools that can be used by producers to assist with manure spreading decisions that protect water quality. This is one of many practical tools that incorporate weather forecasts to plan daily hauling activities for specific locations. http://www.manureadvisorysystem.wi.gov

- Texas A&M University's Texas Animal Manure Management Issues Website- This is an information clearinghouse providing educational materials on regulations and policies and up to date research on animal waste management and animal water quality issues.
http://tammi.tamu.edu/index.html